INFRARED AND SUBMILLIMETER SPACE MISSIONS IN THE COMING DECADE

INFRARED AND SUBMILLIMETER SPACE MISSIONS IN THE COMING DECADE:

PROGRAMMES, PROGRAMMATICS, AND TECHNOLOGY

Edited by

HARLEY A. THRONSON, JR.

Wyoming Infrared Observatory,
University of Wyoming,
Laramie, U.S.A.

and

MARC SAUVAGE, PASCAL GALLAIS and LAURENT VIGROUX

DAPNIA/Service d'Astrophysique,
C.E. Saclay,
Gif-sur-Yvette, France

Reprinted from *Space Science Reviews*, Vol. 74, Nos. 1–2, 1995.

SPRINGER SCIENCE+BUSINESS MEDIA, B.V.

Library of Congress Cataloging-in-Publication Data

Infrared and submillimeter space missions in the coming decade : programmes, programmatics, and technology / edited by Harley A. Thronson ... [et al.].

p. cm.

ISBN 978-94-010-4162-1 ISBN 978-94-011-0363-3 (eBook)

DOI 10.1007/978-94-011-0363-3

1. Infrared astronomy--Congresses. 2. Submillimeter astronomy--Congresses. 3. Astronautics in astronomy--Congresses. I. Thronson, Harley A.

QB470.A1I49 1995

522 68--dc20 95-30187

ISBN 978-94-010-4162-1

Printed on acid-free paper

SPACE SCIENCE REVIEWS / *Vol. 74 Nos. 1/2 October 1995*

List of Participants vii

Foreword xi

E. LELLOUCH / Solar System Observations and Future Missions 1–7

A. FRANCESCHINI, G. DE ZOTTI and P. MAZZEI / Extragalactic Observations and Future Missions 9–16

G.H. RIEKE, E.T. YOUNG and T.N. GAUTIER / Detection Limits in the Far Infrared 17–25

J.-M. LAMARRE, F.-X. DÉSERT and T. KIRCHNER / Background Limited Infrared and Submillimeter Instruments 27–36

F.R. BOUCHET, R. GISPERT, N. AGHANIM, J.R. BOND, A. DE LUCA, E. HIVON and B. MAFFEI / Simulations of the Microwave Sky and of Its "Observations" 37–43

T.G. HAWARDEN, R. CRANE, H.A. THRONSON, JR., A.J. PENNY, A.H. ORLOWSKA AND T.W. BRADSHAW / Radiative and Hybrid Cooling of Infrared Space Telescopes 45–56

M.F. KESSLER / The Infrared Space Observatory (ISO) 57–65

C. SINGER / The Infrared Space Observatory: Telescope Design 67–72

T. MATSUMOTO / Infrared Telescope in Space : IRTS 73–79

S.D. PRICE / Infrared Astronomy on the Midcourse Space Experiment 81–87

P.B. HACKING and M.W. WERNER / The Wide-Field Infrared Explorer (WIRE) 89

E.F. ERICKSON / SOFIA: The Next Generation Airborne Observatory 91–100

P.Y. BELY, H.C. FORD, R. BURG, L. PETRO, R. WHITE and J. BALLY / POST: Polar Stratospheric Telescope 101–112

T. MATSUMOTO / Infrared Imaging Surveyor: IRIS 113–117

R. HILLS / Optics of the Far Infra-Red and Submillimetre Space Telescope 119–123

M.W. WERNER and L.L. SIMMONS / SIRTF: The Moderate Mission 125–138

H.A. THRONSON, JR., T.G. HAWARDEN, A.J. PENNY, L. VIGROUX and G. SHOLOMITSKII / The Edison Infrared Space Observatory 139–144

A. LANGE, P. DE BERNARDIS, M. DE PETRIS, S. MASI, F. MELCHIORRI, E. AQUILINI, L. MARTINIS, F. SCARAMUZZI, B. MELCHIORRI, A. BOSCALERI, G. ROMEO, J. BOCK, Z. CHEN, M. DEVLIN, M. GERVASI, V. HRISTOV, P. MAUSKOPF, D. OSGOOD, P. RICHARDS, P. ADE and M. GRIFFIN / The Boomerang Experiment 145–150

A.E. LANGE, J.J. BOCK and P. MASON / The Far Infrared Explorer (FIRE) 151–156

F.-X. DÉSERT / OLBERS: An Interplanetary Probe to Study Visible and Infrared Diffuse Backgrounds 157–162

A. LÉGER, J.-L. PUGET, J.M. MARIOTTI, D. ROUAN and J. SCHNEIDER / How to Evidence Primitive Life on an Exo-Planet? – The DARWIN Project 163–169

J. SUN / The Near Infrared High Resolution Imaging Camera Program of Beijing Observatory 171–173

R. BLAKE and B.W. JONES / The Effect of Realistic Surface Properties on Low Temperature Space Observatories 175–179

S.M. POMPEA / The Management of Stray Radiation Issues in Space Optical Systems 181–193

A. RAVEX, L. DUBAND and P. ROLLAND / Pulse Tube Refrigerators 195–203

T.W. BRADSHAW and A.H. ORLOWSKA / The Use of Closed Cycle Coolers on Space Based Observatories 205–213

I.D. HEPBURN, I. DAVENPORT and A. SMITH / Adiabatic Demagnetisation Refrigerators for Future Sub-Millimetre Space Missions 215–223

F. SIBILLE / Infrared Detection, Review of Existing and Future Devices 225–228

J.J. BOCK, D. CHEN, P.D. MAUSKOPF and A.E. LANGE / A Novel Bolometer for Infrared and Millimeter-Wave Astrophysics 229–235

LIST OF PARTICIPANTS

Luc AUDAIRE	CEN Grenoble, F
Pierre BELY	Space Telescope Science Institute, USA
Robert BLAKE	The Open University, UK
James J. BOCK	U.C. Berkeley, USA
Franois R. BOUCHET	Institut d'Astrophysique de Paris, F
Olivier BOULADE	SAp-Saclay, CEA, F
Franois BOULANGER	Institut d'Astrophysique Spatiale, F
Thomas W. BRADSHAW	Rutherford Appleton Laboratory, UK
Ezio BUSSOLETTI	Istituto di Fisica Sperimentale, I
Catherine CESARSKY	SAp-Saclay, CEA, F
Giorgio DALL'OGLIO	Universita di Roma, I
Catherine De BERGH	Observatoire de Meudon - DESPA, F
Paolo DE BERNARDIS	Universita "La Sapienza" Roma, I
Thijs DE GRAAUW	SRON-Groninger, NL
Teije De JONG	SRON, NL
Franois-Xavier DÉSERT	Institut d'Astrophysique Spatiale, F
Mark DRAGOVAN	Princeton, USA
Lionel DUBAND	CENG-DRFMC-SBT, F
Roger J. EMERY	Daresbury & Rutherford Appleton Labs., UK
Pierre ENCRENAZ	Observatoire de Paris - DEMIRM, F
Ed ERICKSON	NASA Ames Research Center, USA
Damien FEGER	Cryotechnologies S.A., F
Jean-Jacques FERMÉ	Société Européenne de Systèmes Optiques, F
Alberto FRANCESCHINI	Osservatorio Astronomico di Padova, I
Pascal GALLAIS	SAp-Saclay, CEA, F
Reinhard GENZEL	MPI für Extraterrestrische Physik, D
Jane GREGORIO-HETEM	SAp-Saclay, CEA, F
Robert GRIGNON	Thomson-CSF/RCM, F
Tim HAWARDEN	Joint Astronomy Center, Hawaii, USA
Ian HEPBURN	Mullard Space Science Laboratory, UK
Hannibal HETEM	SAp-Saclay, CEA & ENS, F
Richard HILLS	Cavendish Laboratory, UK
Alois HIMMES	German Space Agency DARA, D
Danièle IMBAULT	SAp-Saclay, CEA, F
Bernard JACQUEMIN	SAp-Saclay, CEA, F
Barrie W. JONES	The Open University, UK
Martine JOUBERT	Laboratoire d'Astronomie Spatiale-Marseille, F
Martin KESSLER	Astrophysics Division, ESTEC/SAI, NL
Lydie KOCH-MIRAMOND	SAp-Saclay, CEA, F
Ernst KREYSA	Max Planck Institut für Radioastronomie, D

Space Science Reviews **74**: vii–ix, 1995.

Jean-Louis LAFON	Société Bertin, F
Jean-Miche LAMARRE	Institut d'Astrophysique Spatiale, F
Philippe LAVOCAT	DAPNIA/SED, CEA, F
Alain LÉGER	Institut d'Astrophysique Spatiale, F
Gérard LELIEVRE	DASGAL-Observatoire de Paris, F
Emmanuel LELLOUCH	Observatoire de Paris-Meudon, F
Reno MANDOLESI	TESRE - BOLOGNA, I
T. MATSUMOTO	Nagoya University, J
Francesco MELCHIORRI	Universita "La Sapienza", I
Theodore L. MILLER	Eastman Kodak Co. Rochester, NY, USA
Bianca OLIVO MELCHIORRI	Atmosphere Physical Institute-CNR, I
Anna ORLOWSKA	Rutherford Appleton Laboratory, UK
Renaud PAPOULAR	SAp-Saclay, CEA, F
Alan J. PENNY	Rutherford Appleton Laboratory, UK
Michel PÉRAULT	École Normale Supérieure, F
Stephen POMPEA	S.M. Pompea & Associates, Tucson, USA
Dominique POULIQUEN	Laboratoire d'Astronomie Spatiale, F
Stephan PRICE	Phillips Lab/GPOB, Hanscom, USA
Jean-Loup PUGET	IAS-Orsay, F
Alain RAVEX	Service des Basses Températures CENG, F
George RIEKE	University of Arizona, USA
Yvon RIO	SAp-Saclay, CEA, F
Louis RODRIGUEZ	SAp-Saclay, CEA, F
Hans-Peter RÖSER	MPI für Radioastronomie BONN, D
Hervé SAINCT	Aerospatiale-Cannes, F
Morvan SALEZ	Jet Propulsion Laboratory, Pasadena, USA
Paolo SARACENO	IFSI-CNR Frascati, I
Marc SAUVAGE	SAp-Saclay, CEA, F
F. SCARAMUZZI	ENEA, Frascati, I
Stephen SCULL	British Aerospace Space Systems Ltd, UK
Guy SERRA	CESR - Toulouse, F
Gennadii SHOLOMITSKII	Space Research Institute-Moscow, CIS
Franois SIBILLE	Observatoire de Lyon, F
Larry SIMMONS	Jet Propulsion Laboratory, Pasadena, USA
Christian SINGER	Aerospatiale-Cannes, F
Bernard SOUCAIL	THOMSON TRT Défense, F
Aimé SOUTOUL	SAp-Saclay,CEA, F
Jinghao SUN	Max-Planck Institut für Astronomie, D
Harley THRONSON	Wyoming Infrared Observatory, Laramie, USA
Thierry TOURRETTE	SAp-Saclay, CEA, F
Thuan TRINH XUAN	SAp-Saclay, CEA, F & Univ. of Virginia, USA
Laurent VIGROUX	SAp-Saclay, CEA, F
Sergio VOLONTE	ESA Paris, F

Helen WALKER	SERC-Rutherford Appleton Laboratory, UK
Michael WERNER	IPAC - Caltech, USA
Paul WESSELIUS	SRON-Groningen, F

FOREWORD

In the last 15 years, an unprecedented revolution in observational astronomy took place. The emergence of CCDs, an almost perfect detector in the visible, has been the first breakthrough. Developments of space technologies have opened new wavelength windows. The Hubble Space Telescope and IRAS have demonstrated the scientific capabilities of space observatories. A similar revolution is still ahead in infrared and submillimeter astronomy. There is certainly no wavelength range which has, over the past several years, seen such impressive advances in technology: large-area detector arrays, new designs for cooling in space, lightweight mirror technologies, and a variety of increasingly powerful groundbased and airborne instruments. Scientific cases for observing the cold Universe are outstanding. The answers to such fundamental questions as what is the primordial fluctuations' spectrum, how do primeval galaxies look like, or what are the first stages of star formation will be provided by observations in the FIR/Submm range. These major scientific issues, together with the new technology available, have triggered a very large development of new programs. As atmospheric absorption and emission are the major limitations to observations at these wavelengths, most of these programs are space missions.

Central to international astronomy in this decade will be the European Space Agency's *Infrared Space Observatory* (ISO), which will be the first true space observatory at these wavelengths. So important will this mission be to astrophysics, that many critical research areas are likely to mark their understanding of the cosmos simply as "Pre-ISO" and "Post-ISO." In addition, European contribution is central to NASA's Stratospheric Observatory for Infrared Astronomy (SOFIA), due to be flying by 2001, and about a decade from now, ESA will launch the first sub-millimeter space observatory, the *Far-Infrared and Sub-Millimeter Space Telescope* (FIRST). Farther into the future, ESA has begun advanced technical studies which may lead to larger-aperture IR space missions (the *Edison* proposal) and an infrared interferometer (currently known as DARWIN) with the specific goal of spectroscopy of extra-solar Earth-like planets.

About three years ago in the US, the National Academy of Sciences Decade Review of Astronomy & Astrophysics ("The Bahcall Report") declared the 1990s to be "The Decade of the Infrared." This was the first time in the history of American decade reviews that a particular wavelength range had been clearly identified as the paramount goal for future exploration. Most of the effort in the US on future IR missions has been in support of SOFIA and of the *Space Infrared Telescope Facility* (SIRTF), which is currently planned for launch early in the next decade. This will be the next observatory-class space mission after ISO and will carry advanced IR detector systems.

The Japan astronomy program has also an important infrared component. The next step is a shuttle launch mission, IRTS. At about the same time than SIRTF, the Japanese are expected to launch the *Infrared Imaging Surveyor* (IRIS), perhaps in collaboration with NASA.

Thus, most of the major space agencies are pursuing ambitious programs to understand the Universe in ways that are only possible at infrared wavelengths. Furthermore, the major programs are becoming more international, if not from the start, then sometimes out of necessity due to increased financial pressures. With an ambitious future, it is important to collect in one place up-to-date descriptions of the missions and the technologies of infrared and sub-millimeter stratospheric and space astronomy. Exchange of information will help to avoid duplications of expensive programs, and might be the basis for future international collaborations.

As this meeting was being planned, the European Space Agency was undergoing its future planning program, Horizon 2000 Plus, which would guide the agency for the rest of the professional lives of most living space scientists and engineers. Given the importance that ESA has played in developing astronomy at these wavelengths, it was appropriate that our meeting be held near Paris. As a major contributor to the ISO instruments, and with a strong interest in the future infrared an submillimeter space missions, the Service d'Astrophysique in Saclay was very pleased to host this meeting. We hope that it will confirm that the years 2000 should be the Infrared Years.

We appreciate the support of Rutherford Appleton Laboratory and Dr Michiel Kolman at Kluwer Academic Publishers.

Laurent Vigroux, Pascal Gallais, Marc Sauvage
Service d'Astrophysique
Gif-sur-Yvette, France

Harley Thronson, Jr.
Wyoming Infrared Observatory
Laramie, Wyoming USA

October, 1994

SOLAR SYSTEM OBSERVATIONS AND FUTURE MISSIONS

EMMANUEL LELLOUCH
Observatoire de Paris, 92195 Meudon, France

Abstract. This paper aims at placing infrared (IR) and submillimeter (submm) observations of the Solar Sytem in the context of future space missions. First, the information that mm, submm and IR observations bring on planets, satellites and comets is reviewed. Then, some lines of future research in this field are explored, and a number of observations that could enhance our understanding of Solar System objects are suggested. Finally, the adequacy of future space missions in this respect is discussed. The specific cases of ISO, FIRST and the proposed Edison are considered.

1. What do we learn from mm, submm, and IR Studies of the Solar System?

1.1. Planets

The IR spectrum of planets consists of two components: (i) the thermal emission (corresponding to the Planck emission of the atmospheres and the surfaces), that peaks in the mid-to-far infrared (15-100 microns) (ii) the solar radiation, reflected at the surfaces or at cloud layers, which peaks near 0.5 μm. In the thermal component, the radiative transfer equation in an atmosphere shows that an observation at a frequency ν probes the atmosphere near the pressure level where the weighting function (defined at the derivative of the transmission with respect to the log of the pressure) maximizes. Therefore, multi-wavelength observations in the thermal component allow a vertical probing of the atmosphere. Information can be obtained on several physical parameters: (i) the gaseous global abundances, from which elemental and isotope abundances can be inferred, constraining the origin and evolution of the planetary atmospheres, (ii) the vertical distribution of gases and of temperature profiles, and their spatial variability, providing information of the physical processes (e.g. photochemistry, dynamics, meteorology...) which govern the atmospheres, (iii) the surface properties: if the atmosphere is not totally opaque and the surface is probed, its emissivity properties can be measured and some information on the surface composition obtained. In the solar reflected component, the radiative transfer equation simply expresses the fact that the solar flux is attenuated by the atmospheric opacity before and after being reflected, either by the surface or by a cloud layer. At the (shorter) wavelengths involved, this radiation is also sensitive to scattering by clouds or aerosols. Spectroscopy in this spectral range provides information on the integrated abundances of gases (unlike in the thermal range, vertical resolution is not possible), the scattering properties of the atmospheres (clouds and aerosol phase function, particle size and con-

Space Science Reviews **74**: 1–7, 1995.

centration), and the composition of surfaces. The latter aspect is particularly important for airless bodies (most satellites, small planets, asteroids).

1.2. Comets

The IR spectrum of comets consists of spectral features superimposed on a continuum due to dust. Below 3 μm, the continuum corresponds to the reflection of the solar radiation by the dust. Above 3 μm, the continuum is formed by the thermal emission of the dust, and depends of course strongly on the dust temperature (i.e. on the heliocentric distance). The spectral features observed in comets can be due to the dust itself (e.g. the silicate emission at 10 μm) or to some radicals in the near IR (e.g. CN, C_2), but most of them belong to parent molecules. So far, 7 parent molecules have been identified in the IR, submm or mm spectrum of comets: H_2O (at 2.7 μm), CO (at 4.7 μm), CH_3OH (at 3.52 μm, and 0.8-2 mm), H_2CO (at 3.59 μm and 0.8-1.3 mm), CO_2 (at 4.25 μm), HCN (at 0.8-3 mm), H_2S (at 1.4-2 mm). In addition, OCS has been tentatively identified at 4.85 μm, and a broad emission near 3.4 μm is attributed to the C-H stretching, although its exact origin remains an open question. Most of the IR detections were obtained in comet P/Halley, but several other comets (P/ Brorsen-Metcalf, Levy, Austin, P/Swift-Tuttle) have been succesfully observed, particularly at mm/submm wavelengths. Besides the information it brings on the dust composition, the IR and (sub)mm spectrum is therefore extremely powerful to study parent molecules. It allows to measure their production rates and to estimate their relative abundances, providing clues for the origin of comets. An example in the detection of H_2S, a very volatile compound, which implies that comets formed and remain stored at very low temperatures. Secondly, some phyical parameters can be reached, among other (i) the spatial distribution of molecules in the coma (e.g. H_2CO seems to have a distributed source), (ii) the rotational temperatures (done so far for water and methanol), (iii) the coma kinematics and the nucleus inhomogeneities (which can be studied at heterodyne resolution from line widths and shifts). Finally, the comparison of the relative abundance inferred in different comets suggests that comets are not all alike.

2. Future Observations

In spite of significant progress in our knowledge of Solar System objects in the last decade, there are still many open questions. We here review some important observations that we feel should be performed in the next future, and we briefly discuss the physical interest of these observations.

2.1. Atmospheres

2.1.1. Global Composition

Much could be learned about the global composition of planetary atmospheres by studying the so far unexplored regions of their spectrum (Encrenaz, 1992). As the highest priority, the submm spectrum (50-1000 microns) of the Giant Planets must be investigated at high spectral resolution (Bézard et al., 1986). First, this range contains in particular two HD lines (at 56 and 112 μm) which, if observed, would provide straightforward measurements of the D/H ratio. The comparison of the D/H ratio in the four Giant Planets would give insight on the nature of the two deuterium reservoirs that apparently were present at the time of Solar System formation. Second, this range contains rotational transitions of many so-called disequilibrium species. These are those molecules which are thermodynamically unstable in the upper Giant Planets tropospheres but are nevertheless observed there because they are transported upwards by strong internal convection. Three of them have been detected (CO, PH_3, AsH_3) and many other have transitions in the submm range, notably the halides (HCl, HF, HI, HBr) and H_2Se. Detecting some of them would provide new elemental abundances and information on the vertical transport. Another very important observation would be the recording of the full far IR and submm spectrum of Titan (Coustenis, 1992). This would allow in particular searching for water (through its rotational lines) and for several complex hydrocarbons and nitriles (through their vibrational bands), thereby investing two basic questions, namely the nature of the oxygen chemistry (origin of CO, CO_2?) and that of the organic and prebiotic chemistry (how evolved is it?). Finally, it would be useful to investigate the regions of weak flux, in particular the 3-7 μm spectrum of Titan, Uranus and Neptune. This would allow for a search of PH_3 in Uranus and Neptune, and study the possible presence of a C-H feature near 3.4 μm in these bodies.

2.1.2. Spatial and Temporal Variations

Spatial and temporal variations in the composition and temperature structure of the planets are generally indicative of dynamical and/or chemical effects. In many cases, the mechanisms controlling the variations, however, are not established in details. Without trying to be exhaustive, a few examples can be given (i) on Venus, the diurnal variations of the mesospheric CO and O_2 seem to be controlled by the general circulation, (ii) on Mars, the entire photochemistry involves complex relationships between many species (water and ozone in particular) and with the aerosol distribution, (iii) on Titan, most photochemical species exhibit spatial and seasonal variability (iv) on Jupiter, PH_3 seems to be enhanced at the North Pole, is it a result of a latitudinal variability of the deep convection, do other species behave similarly? In all these examples, it is clear that the most efficient way to address the problem is through a coordinated monitoring of many species, that can be best achieved at infrared and submm wavelengths.

2.1.3. Search for Tenuous Atmospheres

Another area is the search for tenuous atmospheres. In particular, while Pluto and Triton do have atmospheres, they were never spectroscopically detected in the IR. The search for CO and CH_4 should be conducted. Finally, the presence of atmospheres on other bodies must be investigated, e.g. on the Galilean satellites.

2.2. SURFACES

The composition of planetary surfaces, either rocky or icy, are still not well characterized. Regarding icy surfaces, some examples of important problems to be solved with the help of infrared spectroscopy are (i) what is the exact composition of the martian polar caps (CO_2 vs. H_2O)? (ii) are there any other ices (to be searched for at $\lambda \geq 2.5\ \mu$m) on Pluto and Triton besides the detected N_2, CO, CH_4 and CO_2 ? How will Pluto's surface composition evolve after perihelion? (iii) What is the horizontal, vertical and temperature distribution of the SO_2 frost on Io? Concerning rocky surfaces, the entire IR spectrum must be studied at moderate resolution, to determine the composition of small satellites, rings, asteroids and Kuiper Belts objects.

2.3. COMETS

Future lines of investigation of comets are concerned with (i) their composition, (ii) the coma physics, and (iii) the monitoring of cometary activity (Crovisier 1992).

2.3.1. Cometary Composition

It is desirable to improve our knowledge of cometary composition, both of the dust and of the gaseous component. An important question regarding the dust is to know whether other compounds besides silicates are present: carbonates have been tentatively identified near 6.8 μm, and some emissions seem to be present at 24-28 μm, but clearly further observations are necessary. The most outstanding questions, though, are related to the volatile parent molecules. What are the progenitors of the C_2 and C_3 radicals (methane has not been detected) Do comets contain hydrocarbons, ammonia? What is the exact origin of the 3.2-3.6 μm emission? Various answers have been given to this last question. They diversely invoke methanol, a combination of saturated and unsaturated hydrocarbons, large PAH-like molecules, and grains with organic mantles. Clearly, higher resolution observations of this emission are necessary. Also, exploring the difficult 6 and 10 μm ranges where some of the candidates have features should help discriminating between the various possibilities. A question related to cometary composition is that of the diversity of comets. It is now necessary to build a database of comets to study the variations in their composition.

2.3.2. Coma Physics

Heterodyne and IR observations have shown that molecular transitions can be used as tracer of coma physics. In particular, the temperature and the density in the coma can be mapped. The distribution of temperature as a function of distance from the nucleus can be deduced from rotational distributions. For this task, the better tracers will be the water lines, either at 2.7 μm (already used), or in the submm range (557 and 762 GHz).

2.3.3. Monitoring the Cometary Activity

A very important tool to understand the origin and nature of comets is the monitoring of their activity. On a short term, using the nucleus rotation, the variations in activity reflect the nucleus inhomegeneities, providing clues to the origin of cometary nuclei. On a long term, it is important to follow activity over large distances, to monitor outbursts. The activity of comets at large heliocentric distances is a new and basic subject, since the discovery of CO on comet P/Schassmann–Wachmann 1 (IAU Circ. 5929). This observation demonstrates that cometary activity is not necessarily driven by H_2O. Understanding the large distance activity is important for at least two purposes (i) a correct retrieval of nucleus abundances from parent molecules abundances measured in the coma close to the Sun, (ii) understanding the relations between comets and asteroids (for example, 2060 Chiron, an object classified as an asteroid, shows some activity at 10 AU).

3. Adequacy of Future Space Missions

In the light of the previous suggested observations, we here discuss the possibilities that will be offered by future space missions. As benchmarks, we examine the capabilities of ISO, FIRST and the proposed Edison.

From the point of view of Solar System studies, ISO is an excellent observatory thanks to its high sensitivity and high spectral resolution. On the other hand, it essentially offers no spectroscopic spatial resolution on planets. ISO is also limited by its relatively short lifetime and the limit in sky coverage is a problem for the study of comets. The advantage that FIRST will bring is four–fold: (i) the heterodyne resolution at 500-1000 GHz, (ii) an increase sensitivity at 100-200 μm, (iii) some modest spatial resolution (7" at 100 μm), (iv) a longer lifetime. Finally, Edison, as a proposed project, is still not defined in detail, but for our purpose, we will assume it to be 5–10 times more sensitive than ISO and to have a lifetime 5–10 times longer.

3.1. ATMOSPHERES

ISO is an excellent instrument for any studies of Mars, Jupiter and Saturn. The entire IR spectrum of these objects will be easily recorded at high S/N ratio. ISO is also very good for Uranus, Neptune and Titan, except in the regions of minimum flux (3-7 μm). We therefore anticipate a large progress in our knowledge of global atmosphere composition. FIRST, for its part, will permit, thanks to the heterodyne spectroscopy, to address specific problems such as the monitoring of water and O_2 on Venus and Mars, or the search of ammonia in Titan and Neptune. FIRST should also allow the mapping of disequilibrium species on Jupiter, e.g. PH_3 at 100 μm. Finally, a project like Edison would be well suited to fill the regions of weak flux that are difficult for ISO. Its long lifetime would also allow to monitor the entire IR spectrum of Mars over several martian years.

3.1.1. Surfaces

ISO will permit to record the mid-IR spectrum of the Galilean satellites (especially Io) and of a few (10-15) bright asteroids of several classes. It could also allow a measurement of Pluto's surface temperature. While FIRST will not be suited to study surfaces, the expected highest sensitivity of Edison will make it unvaluable for the study of weak objects. Mid-resolution spectroscopy of Saturn's and Uranus' satellites and of many asteroids (allowing to build a database) will be possible. Pluto's surface and its evolution with time may be also monitored. Finally low resolution spectroscopy might be possible on Kuiper Belt objects.

3.1.2. Comets

A breakthrough in our knowledge of comets is expected with the operation of ISO. The entire IR spectrum will be recorded on 1 or 2 comets (current targets are P/Kopff and P/Wild-2) and high S/N ratio investigations of some specific ranges (2.6-2.9 μm for H_2O and 4-4.8 μm for CO and CO_2) will be performed, allowing to define a reference spectrum. In addition, photometric studies of the onset of activy on a distant comet and low resolution gas/dust mapping will be also possible. FIRST will permit another qualitative step, since it will permit to resolve cometary lines, and in particular to monitor the activity in the water 557 and 762 GHz lines. This will allow to follow the evolution with time of several (even weak) comets, and to study the coma kinematics using line shape. In addition, high resolution searches of other parent molecules (NH_3), water isotopes and water ions will be possible. Finally, Edison will take advantage of its increased sensitivity and lifetime (augmenting the chance to observe a fresh bright comet) to (i) construct a high resolution reference spectrum in the entire IR, (ii) study the diversity of comets, and (iii) study cometary activity at large heliocentric distances.

Acknowledgement

I am indebted to J. Crovisier for his help on the cometary part of this review

References

Bézard, B., Gautier, D., Marten, A.: 1986. *Astron. Astrophys.* **161**, 387.

Coustenis, A.: 1992. In, *Infrared Astronomy with ISO*, Eds. Th. Encrenaz and M.F. Kessler, p. 197.

Crovisier, J.: 1992. In *Infrared Astronomy with ISO*, Eds. Th. Encrenaz and M.F. Kessler, p. 221.

Encrenaz T. 1992. In *Infrared Astronomy with ISO*, Eds. Th. Encrenaz and M.F. Kessler, p. 173

EXTRAGALACTIC OBSERVATIONS AND FUTURE MISSIONS

A. FRANCESCHINI , G. DE ZOTTI and P. MAZZEI
Osservatorio Astronomico di Padova
Vicolo Osservatorio 5 – I-35122 Padova – Italy

Abstract.
We briefly review some questions of extragalactic astrophysics and cosmology that would most benefit from future missions outside the Earth's atmosphere in the IR and sub-millimeter. These include the formation and early evolution phases in galaxies and the probably related question of quasar formation; the observation of Active Galactic Nuclei embedded in thick dusty structures (torii) and its impact on the still debated unified model of AGN activity; the observability of radiation processes occurring at very high z through background measurements; the investigation of the large scale structure and velocity field in the distant universe; and studies of the interstellar medium in galaxies. Some more emphasis is given on the galaxy formation problem, because we believe that IR-mm sensitive observations will be crucial to its final solution.

Key words: IR galaxies – Active Galactic Nuclei – dust emission – galaxy formation

1. Introduction

The IR and sub-millimeter wavelength domain is expected to provide new exciting developments to the extragalactic research of the next decades. This may be easily inferred from consideration of the results already achieved, in spite of the limited resources devoted to the field.

Most of our knowledge of the IR sky comes from the IRAS mission (Soifer et al., 1987). The discovery of the powerful far-IR emission of galaxies in general (and of spirals, starbursts and AGNs in particular) is among its brightest successes. A new class of extremely powerful emitters (the ultra-luminous IRAS galaxies) has been discovered uniquely through their far-IR output (Sanders et al., 1989). The currently best case of high-z primeval galaxy candidate (IRAS 10214+4724, Rowan-Robinson et al., 1993) was found in a sample of faint IRAS sources. All this emphasizes the key role of dust in degrading the optical energy emission and re-shaping the broad-band spectral appearence of galaxies.

Furthermore, deep explorations from ground in a few restricted wave-bands (the K and L near-IR bands, the 10 μm mid-IR, and a few sub-mm windows) also start to be quite informative. Mid-IR observations are getting close to the activity sites in galaxies, such as the star-forming regions and the inner dusty environments of active nuclei. Finally, CO line and continuum observations in the millimeter region are providing information on the interstellar medium in both nearby and very distant objects.

Space Science Reviews **74**: 9–16, 1995.

We will devote this review to an educated extrapolation of what is currently known to argue about observational prospects of future missions.

2. Formation and early evolution phases in galaxies

How did galaxies form and evolve is still a mystery to a large extent. Important constraints have been set, however, by deep surveys in various electromagnetic bands. A key result of optical searches (through faint counts and spectroscopic follow-up, narrow band imaging, and searches for spectral emission features) is a failure to reveal any populations of luminous starbursts at high redshifts, corresponding to the main formation epoch. As argued by Franceschini et al. (1994), such inability to detect primeval galaxies – or even objects at $z > 1$ – in the optical band may be understood in two alternative ways. One often discussed possibility is that at such redshifts luminous galaxies simply do not exist, because they formed only more recently (at $z \leq 1$) through accretion of smaller subunits, as envisaged in the *merging* picture of galaxy formation (e.g. Broadhurst et al., 1992).

An alternative possibility is that they were already formed by $z = 1$, but cannot be detected because the large amounts of optical-UV photons emitted by newly born stars are suppressed by dust present in the diffuse medium from which stars are formed. Something similar happens in the local universe, where starforming regions are usually buried in heavily extinguished molecular clouds. In such an event, even a small fraction of the early-produced metals, condensing into dust grains, may yield appreciable extinction and dust re-radiation, during a phase in which an important fraction of the barionic content was still in a gaseous form, such a phase also coinciding with that of major energy production by stellar nucleosynthesis.

Explorations performed at IR wavelengths are then needed to investigate this scenario: not only dust extinction becomes negligible longward of few microns, but dust itself is expected to re-radiate the corresponding energy in the far-IR.

Two complementary approaches may be followed to test these ideas. The basic one will be to survey blindly and thoroughly portions of the sky to the faintest possible flux levels. A sizeable fraction of the entire ISO lifetime, for example, will be devoted to such unbiased surveys with the most sensitive filters in both the mid- and far-IR. Certainly, future IR-mm space missions too will be expected to devote comparable amounts of time to this kind of observations. Figure 1 reports the predicted redshift distributions for complete galaxy samples selected at 90 μm down to the ISO confusion limit, following the two above mentioned formation modes. The curves corresponding to *disk-dominated* (spirals) and *early-type* (E's and S0's) galaxies are indicated. Detailed evolutionary population synthesis models have been discussed by Mazzei et al. (1992, 1994), accounting for most available infor-

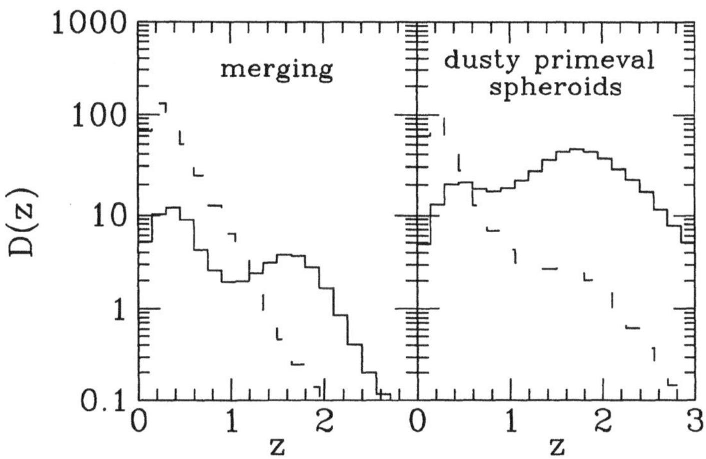

Fig. 1. Redshift distributions of galaxies selected at $\lambda = 90$ μm down to $S_{90} = 10$ mJy over 1 sq. degree ($H_0 = 50$, $q_0 = 0.5$). Dashed lines: ***disk*** galaxies; continuous lines: ***early-types***. Left panel assumes that galaxies form through merging accompanied by luminosity evolution. Disk galaxies are a factor 3 times less massive and 7 more luminous at z=1 than locally, whereas spheroids are 10 times less massive and evolve in luminousity as discussed in the text. Right panel assumes a standard galaxy formation scenario, with formation epoch after 1 Gyr of the big-bang and "dusty" evolution of spheroids.

mation on chemistry, photometry, dust properties, and their change with cosmic time (see Figure 2 below).

The two evolution schemes outlined in Figure 1 do not imply relevant differences for disk galaxies, because their SED's are not expected to strongly evolve with time. On the contrary, one can notice the power of such observations to test different formation patterns for strongly evolving spheroids. Note that the right-hand panel of Figure 1 corresponds to somewhat extreme recipes for the evolution of the average optical depth of these systems. Such a case would imply that over 50% of all galaxies above the ISO sensitivity limit would be found at $z > 1$! A less dramatic evolution would correspond to fainter far-IR fluxes for high-z starbursting galaxies, which ISO could possibly not detect. A definite answer might not come before the launch of future observatories (EDISON, SIRTF), better optimized for the far-IR.

An alternative approach to that of sky surveys will be to concentrate the efforts on a limited number of candidate primeval galaxies discovered through various observational criteria. An obvious one is to look at the source's energy emission at long wavelengths (radio, mm, or infrared), where dust is ineffective or positively contributes to the flux. Another one is to look at indirect effects, e.g. gravitational lensing of compact radiosources and spectral absorption features in the direction of high-z QSOs. We mention below a few such interesting targets.

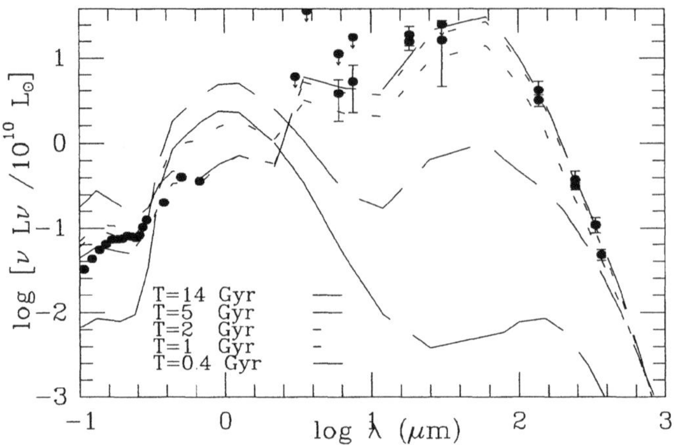

Fig. 2. Broad-band SED of IRAS F10214+4724, a galaxy at $z = 2.28$ found among 1400 faint 60 μm sources (see Rowan-Robinson et al., 1993). It is compared with spectra of "opaque" star-forming spheroids at various ages, according to Mazzei et al. (1994). Best-fit parameters of $t \simeq 1$ Gyr, $L_{\rm bol} \simeq 2.7\ 10^{14}$ $L_\odot$, $M_{\rm barion} \simeq 10^{13}$ $M_\odot$ and $SFR \simeq 3\ 10^4$ $M_\odot$/yr (Mazzei and De Zotti, 1994) are implied.

The most studied one is the galaxy IRAS F10214+4724. In Figure 2 a collection of data on the source is compared with predictions of an "opaque" starburst model for various galactic ages. It is seen that a spectrum corresponding to an age of 1 Gyr nicely describes the data for over 4 decades in wavelength. Two main problems about such a unique source concern the origin of the huge bolometric emission (if dominated by a starburst or significantly contributed by nuclear activity of a collapsed object; see Elston et al., 1994), and the true object's luminosity (the abnormally high power might be caused by gravitational lensing). If indeed its intrinsic luminosity is much smaller because of lensing, this would have implications on the observability of such evolution phases in primordial objects, suggesting that it could be quite marginal for ISO, and rather require more sensitive experiments.

Radio selection has provided rich lists of high-z galaxies, whose luminosity is mostly produced by non-stellar nuclear activity. A few of them, however, display also properties of forming primeval galaxies (hereafter PG). Figure 3 reports optical-mm data for a few such candidates, compared with typical redshifted SED's of AGNs and of "dusty" starbursting galaxies. The lack of information in the whole IR domain prevents us to evaluate the relative importance of these two contributions and to establish if PG's are hidden, or not, among high-z radio galaxies.

Indirect evidences of the existence of massive, dusty objects at substantial redshifts should be found also in the optical. A way is to spectroscopically survey the redshift space in the line-of-sight to a distant quasar. Spectra of virtually all high-z QSO show the presence of at least one damped Lyα

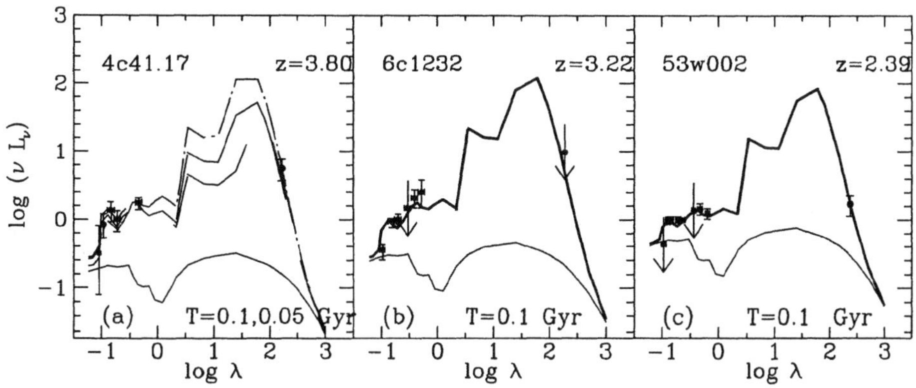

Fig. 3. Broad-band SED's of three candidate PG's high redshift radio galaxies. Thin line is the average optical-IR spectrum of local AGNs (Granato and Danese, 1994), while the thick line includes the contribution (dominant in the far-IR) of "dusty" early spheroids.

system. The metal abundance and dust/gas ratio inferred from extinction effects on the background QSO are typically 1/10 solar (Pettini et al., 1994), implying that these systems are probably forming disks of spirals. On the other hand, heavily enriched forming spheroids on the line-of-sight obscure the background QSO itself, so there may be an anti-bias between the detection of a distant QSO and the presence of a dusty PG along the view. One damped Lyα (in QSO2206-199), however, seems to have solar abundance of both metals and dust, while another one (QSO0528-250) to contain a huge amount of molecular gas (Brown & van den Bout, 1992). IR observations would be needed to clarify the nature of these objects.

Finally, the existence of massive compact opaque objects at high z may be checked by looking at their gravitational lensing effect on background objects. Again, if optical selection may be unfavoured, radio surveys would be ideal for the discovery of dusty gravitational lenses. Larkin et al. (1994) showed that dust probably has a major effect on at least two of the 13 well-established radio-selected lens systems (0414+0534, 1131+0456).

These observations – that could only be made from space, as they require orders-of-magnidutes improvements in sensitivity with respect to current preformances on ground even in the mid-IR atmospheric windows – will contribute to provide not only a decisive test of galaxy formation scenarios, but also information on some peculiar objects in the distant universe.

3. Quasar formation. Hidden AGNs. The unified AGN model.

Millimeter continuum observations have shown that at least a few high-z optical QSOs (BR1033-0327, BR1202-0725) are possibly embedded in massive envelopes of enriched gas (Isaak et al., 1994). Measurements from space of the overall IR continuum and of IR spectral lines will allow an accurate modelling of dust emission and of the ISM in such objects.

These investigations will be particularly relevant to test the idea that quasars trace the sites of formation of the inner dense regions of spheroids (e.g. Terlevich and Boyle, 1994). According to this view, bright optical emission would correspond to a late phase in the formation process, while earlier ones should be characterized by heavy extinction from a dusty surrounding medium. IR observations would then be needed to appropriately select candidates and follow them up into the greatest detail.

Even for low-z AGNs circum-nuclear dust seems to play an important role in re-processing the optical-UV emission in both continuum and lines. The optical spectral-line appearence and the various spectral types (Sey 1 and 2) are interpreted as due to the filtering action of a dust torus present in the inner regions. IR bands, also in connection with the X-ray ones, are then expected to contain the complementary information with respect to the optical to test this unified picture of AGN activity. ISO will certainly allow a lot of progress, but better spatial resolution will be needed to make the final statements.

4. Background light and radiation processes at very high redshifts.

Most part of the redshift interval after the recombination ($5 < z < 1500$, the so-called *dark* ages) is just matter of speculation, as no direct astronomical information has been obtained from current surveys in any wavebands. It would be of key interest, of course, to at least partially fill in the observational gap between the almost undifferentiated primeval plasma mapped in the microwaves and the well structured universe we see at $z < 5$. It is commonly believed that such a possibility could be made available by observations of the background light in two wavebands, the cosmological windows in the L infrared band and in the sub-mm from 300 to 700 μm. It is not a case that these two coincide with the wavelengths where we expect to observe the redshifted peak of photospheric emission from pregalactic stellar generations and the redshifted peak of dust re-radiation, respectively.

Preliminary inferences about the extragalactic IR background have been made from the observed γ-ray spectrum of extragalactic sources (e.g. Dwek and Slavin, 1994). These seem to rule out dominant contributions from exotic sources as exploding stars, very massive objects and decaying particles. COBE, after a careful subtraction of foreground emissions from the Galaxy and from planetary dust, is expected to soon report on estimates of the integrated IR light, but with no spatial information in practice (by the way, a lack of spatial resolution is probably making difficult the evaluation of the starlight contribution to the background observed by DIRBE in the near-

IR window, and similar problems also possibly hamper FIRAS observations in the sub-mm). Another reason making spatial resolution critical is that any inferences about a truly primeval background (due to $z > 5$ sources) require that the extragalactic background from galaxies is properly subtracted. Again, future missions with good sensitivity and spatial resolution, and low stray-light contamination, will probably be needed to fully characterize radiation processes occurring at very high redshifts.

5. Large-scale structure and motions in the distant universe.

Sky surveys performed at sub-millimeter wavelengths (and to a minor extent also in the far-IR) are expected to yield a totally different view of the sky with respect to optical surveys. The K-correction at some hundreds microns is so large and positive ($\propto [1+z]^{3-4}$) that even at moderately bright fluxes an important fraction of the selected objects would be found at substantial redshifts (Franceschini et al., 1991), whereas even deep optical surveys sample mostly the low-z universe. Note that the contamination by galactic sources is expected to be small at such wavelengths, which are then ideal to probe the large-scale structure and its evolution with cosmic time. Because of K-correction, surveys at increasing wavelength from the far-IR to the mm will provide galaxy samples at increasing average redshift.

Part of the job of surveying large pieces of sky will be done from ground in the few accessible wavebands, but certainly an important role will be played by the planned space missions (FIRST in particular). Large format detector assemblies will be needed to collect rich enough samples.

A fundamental tracer of the large-scale structure of the universe, the clusters of galaxies, will be studied at mm wavelengths through the Sunyaev-Zeldovich effect, whose *thermal* component is due to the scattering of background photons by the hot intergalactic electrons. This distortion is negative in the radio and becomes positive shortward of $\lambda = 1.4$ mm. The *kinetic* SZ effect is a Doppler distortion due to the motion of the cluster with respect to the rest-frame of the background photons, and is proportional to the cluster peculiar velocity. It may be observable close to 1.4 mm, at the minimum of the thermal distortion. A suitable combination of sensitive multi-filter observations from dedicated space missions will allow to measure both the thermal and the kinetic distortions for several hundreds of distant cluster (Puget et al., 1993). This will allow to probe the large-scale velocity field in the distant universe and compare it with local estimates. The largest accessible redshifts will be determined by the cosmic evolution of the plasma parameters (column density, temperature), which is very poorly understood at the moment.

6. The interstellar medium in galaxies.

Some fundamental questions about galaxies, as the total amounts of cold gas and dust, their optical extinction and their contribution to the dynamics of the outer regions (Valentijn, 1994), are still matter of a lot of controversy. The main reasons for this are in the limited spatial resolution, sensitivity and spectral coverage of the IRAS survey data, while sub-mm observations from ground are not yet sensitive enough and in any case confined to a few wavebands. This lack of information also entails that the emissivity properties of large grains at low-temperatures are poorly known.

A substantial progress of our knowledge of dust content in galaxies and its cosmic evolution will be made when the entire far-IR/millimetric spectral continuum will be measured with enough spectral and spatial detail. The latter require a space observatory with a large enough primary collector to pull down as far as possible the most serious limitation of such missions: the diffraction-limited spatial resolution.

Even more crucial is the requirement of a space platform for the study of various kind of molecular (e.g. the H_2, H_2O) and atomic (CII, OI, NII) species, of ices, dust grains and aromatic complexes (PAH) through observations of the IR-mm spectral lines and band features for galaxies up to relatively large distances (several hundreds Mpc for the brightest of them).

All this will allow to relate the birth and death of stars, a crucial step towards a physical understanding of cosmic evolution of galaxies.

References

Broadhurst, T.J., et al.: 1992, *Nature* **355**, 55.
Brown, R., and vanden Bout, P.: 1992, *ApJ* **397**, L19.
Dwek, E., and Slavin, J.: 1994, *ApJ*, in press.
Elston, R., et al.: 1994, *AJ* **107**, 910.
Franceschini, A., Mazzei, P., De Zotti, G. and Danese, L: 1994, *ApJ* **422**, 81
Franceschini, A., Toffolatti, L., Mazzei, P., Danese, L. and De Zotti, G.: 1991, *A&AS* **89**, 285.
Granato, G., and Danese, L.: 1994, *MNRAS* **268**, 235.
Isaak, K.G., McMahon, R.G., Hills, R.E. and Withington, S.: 1994, *MNRAS* **269**, L28.
Larkin, J.E. et al.: 1994, *ApJ* **420**, L9.
Mazzei, P. and De Zotti, G.: 1994, *MNRAS* **266**, L5.
Mazzei, P., Xu, C. and De Zotti, G.: 1992, *A&A* **256**, 45
Mazzei, P., De Zotti, G. and Xu, C.: 1994, *ApJ* **422**, 81
Pettini, M., et al.: 1994 *ApJ* **426**, 79.
Puget, J.L., et al.: 1993, Proposal for M3 medium-size ESA project.
Rowan-Robinson, M., et al.: 1993, *MNRAS* **261**, 513.
Sanders, D. et al.: 1989, *ApJ* **347**, 29.
Soifer, B., Houck, J. and Neugebauer, G., 1987, *ARAA* **25**, 187.
Terlevich, R., and Boyle, B.: 1993, *MNRAS* **262**, 491.
Valentijn, E.A.: 1994, *MNRAS* **266**, 614.

DETECTION LIMITS IN THE FAR INFRARED

G. H. RIEKE and E. T. YOUNG
Steward Observatory
University of Arizona
Tucson, AZ. 85721

and

T. N. GAUTIER
Jet Propulsion Laboratory
Pasadena, CA. 91109

Abstract.
The advent of far infrared arrays will change fundamentally the means of analyzing observations in this spectral region. Sources much fainter than traditional "confusion limits" will be extracted from images by using computer algorithms similar to CLEAN or DAOPHOT. We have conducted numerical experiments to evaluate these techniques and show that they will permit long integrations ($\sim$ 10,000 sec at 60 μm, $\sim$ 200 sec at 100 μm) to achieve nearly photon-background-limited performance and hence very deep detection limits. The dominant noise sources – photon noise, confusion by distant galaxies, and confusion by IR cirrus – scale with nearly the same power of the telescope aperture. As a result, the integration times required to reach confusion limits are nearly aperture-independent.

Key words: confusion limit, infrared arrays

1. Introduction

In planning far infrared astronomy missions, it is important to understand the types of noise they will encounter. Confusion by faint galaxies and by infrared cirrus emission can limit the signal to noise achievable in the far infrared. We calculate the magnitude of these effects where the telescope beam is limited purely by diffraction and the confusing galaxies appear as unresolved point sources. These assumptions are appropriate for the majority of envisioned far infrared telescopes, which are designed with image quality and pointing stability required for near infrared operation and which are therefore virtually perfect optical systems in the far infrared.

A number of previous works have argued that natural-background-limited telescopes will be severely limited by confusion noise in a few seconds of integration. These authors have neglected the improvements in performance that will be possible with imaging arrays in the far infrared. When used with arrays in an optimum manner, a telescope cold enough to be photon noise limited from celestial backgrounds has far superior performance to warmer telescopes throughout the far infrared. Telescopes sufficiently cold to be natural background limited can be essentially photon noise limited at wavelengths of 100 μm and shorter for long integrations (e.g., as long as 10,000 sec at 60 μm).

Space Science Reviews **74**: 17–25, 1995.

2. Point Source Confusion Noise

Condon (1974) demonstrates that estimating point source confusion noise requires only knowledge of the integral source density and of the field of view of the instrument, so long as the point source response is taken to be the "effective beam solid angle", Ω_e, which accounts for the wings of the beam. In the regime of interest to us, the integral source counts go approximately as a power law with index $\beta = -1$. In this case, the limiting noise goes as the square of the beam diameter. This steep dependence can lead to substantial variations in estimates of the confusion noise limit if differing assumptions are made about the achievable beam diameter.

IRAS observations provide a direct measure of the far infrared source density. At 60 μm, Hacking & Houck (1987) and Hacking (1994, private communication) find $\sim$ 20 objects deg^{-2} brighter than 50 mJy. Above this flux limit, the integral counts follow a power law as $S^{-3/2}$, as expected for Euclidean space. However, below this flux level the most prevalent galaxies become increasingly more distant, and cosmological corrections will become important so that the counts will deviate from the Euclidean dependence. Therefore, a model of the behavior of galaxies must be constructed from the IRAS data and used with an appropriate cosmology to extrapolate to lower flux densities and determine the density of faint galaxies on the sky. The model we have used will be described elsewhere (Rieke, Young, and Gautier, in preparation); here, we note that the source counts from our model are in good agreement (within a factor of $\sim$ 1.5) with the predictions of Franceschini et al. (1991) and Helou & Beichman (1991). At the flux densities and wavelengths of interest for limiting far infrared measurements with cryogenic observatories, all the counts behave roughly as power laws with $\beta \sim -1$. The predictions are not strongly dependent on cosmological assumptions or on the amount of galaxy evolution when normalized to the 60 μm counts described above.

Given the densities of far infrared galaxies on the sky, to determine the confusion limit of an instrument we need to determine the smallest effective beam area that it can use. For future missions, the effective beam area should be determined using source extraction techniques appropriate to observations with imaging arrays such as will soon be available for the far infrared (Young et al., 1993). In the following, we investigate whether optimal source extraction with oversampled array images will allow useful data to be obtained in crowded fields that might be considered confusion limited with a single detector. We have taken a numerical approach that allows us to simulate the operation of various source extraction procedures using Monte Carlo methods. We have verified that our results agree with the analytic treatment of Condon (1974) where the results can be compared. In our Monte Carlo simulations, we have combined the effects of confusion and photon noise so that the results realistically model the amplification of the photon noise as the extraction technique operates in increasingly crowded fields. In comparing results with varying pixel sizes, we have assumed that the pixels operate at the photon noise background limit. We have also simulated problems where the required signal to noise is modest, so that small errors in determining the beam profile do not affect our results. These latter two assumptions allow us to address a "fundamental" confusion limit that will not depend critically on the assumed parameters for the instrument.

The definition of confusion limit depends on the nature of the investigation. For example, an astronomer making an unbiased survey for infrared-bright galaxies would be less annoyed at being "confusion limited" in such objects than would one wishing to determine the flux density of a quasar in the same set of data. In the following, we have considered only the second type of observation since it yields the most stringent measure of the confusion limit.

In our experiments, we generated artificial data by drawing sources randomly from a power law distribution $\mathrm{N}(> S) = \mathrm{C}\ S^{\beta}$, where S is the flux density and C is a constant. Sources were placed at random positions in a field of 57×57 pixels at an average density of 25 per λ/D diameter beam, where D is the telescope aperture. The faintest sources were therefore roughly only 1% of the flux densities at the achievable detection limits. We left out of this field any sources at more than 1000 times the rms noise level, under the assumption that any deep survey field would be selected to avoid extremely bright sources. Each source was convolved with an Airy pattern of full width at half maximum of 10 pixels and gaussian-distributed random noise was added to each pixel. This noise was scaled with pixel size under the assumption of background limited operation. A test source of known amplitude and with an Airy pattern profile was placed in the center of this "data" array.

Initial tests set $\beta = -1$. We used a source extraction method closely related to the CLEAN algorithm. The data array was deconvolved by identifying the brightest pixel and subtracting an Airy pattern of amplitude 1/3 the rms gaussian noise. The position and amplitude of this subtracted flux contribution were stored in another "deconvolved" array and the procedure was repeated, incrementing the amplitudes in the deconvolved array as small amplitude sources were subtracted from the data array. The subtraction was stopped when the variance in the data array increased with subtraction of an additional small source.

The program then computed the estimated flux density of the test source in the deconvolved array. The sky level was determined as the average of the surface brightness between radii of 1 and 2.5 λ/D, after rejecting all peaks at greater than 3 times the rms noise in this region. This rejection was based on the hypothesis that one would avoid obvious neighboring sources in computing a sky level. The source flux density was determined by integrating the signal within apertures of various sizes and subtracting the sky contribution. 400 Monte Carlo integrations were run for each value of assumed noise. It was found that the optimum extraction procedure in the deconvolved image was to integrate the signal within a sharp-sided angular aperture of diameter 0.5 to 0.7 λ/D. Extractions in this size (and neighboring size) apertures reproduced the input source strengths accurately, with no significant biases toward under- or over-estimation.

The artificial central source was selected so that the final signal to noise (including confusion noise and amplified photon noise) would be near five. This case is appropriate to a deep survey where source detections are achieved down to the limiting noise. For each Monte Carlo run of 400 cases, we determined the rms fluctuations in the estimated brightness of the central source, so the suite of runs yielded the relation between density of confusing sources on the sky and noise in the measurement of the central source. We fitted this relation to the expected theoretical behavior, with the effective beam diameter as a free parameter; an excellent fit was achieved with an assumed beam of angular diameter 0.8 λ/D. That is, in this application and with the CLEAN-like method of source extraction, we would successfully predict the total noise of the system including amplification of the photon shot noise if we assumed we were observing with a sharp sided aperture of diameter 0.8 λ/D. The results were found to be largely independent of the selection of reference field, so long as the inner radius was $\geq 0.6\ \lambda/D$. Our results represent a modest amount of superresolution over the full width at half maximum of 1.03 λ/D for an Airy function.

A second set of simulations considered the effects errors in the determination of the point source response might have on the results reported above. We found that the performance of the CLEAN-like algorithm was unaffected in this application by errors in pixel-to-pixel response of 2%, which is the maximum photometric error observed with the far infrared arrays we are developing (Young et al., 1993). In addition, there was no degradation due to artificial errors in the width of the point source response of up to 10%. Given the modest signal to noise of our test observation and the modest degree of superresolution achieved, these results are not surprising.

A third set of simulations was run with a power law source index of $\beta = -1.5$, with results very similar to those just described. Again, the behavior is adequately described by the simple theory if a beam diameter of 0.8 λ/D is assumed. The experiments described above assumed an extreme degree of oversampling, i.e., pixels of 0.1 the diffraction-limited beam diameter. An additional set of experiments addressed the use of larger pixels. Here, the data frames and CLEAN beam were both measured with pixels respectively of 0.3, 0.5, and 0.7 λ/D. It was assumed that the telescope was substepped to maintain the same sampling interval as before, that is 0.1

λ/D. For example, with the 0.5 λ/D pixels, it would be required that data be taken at 25 telescope pointings on a 5×5 grid with 0.1 λ/D between points. 200 iterations were made for each case, all at the same confusing source density where the previous experiments indicated confusion noise would degrade the photon noise limit by a factor of 3. These experiments produced similar results to those with the 0.1 λ/D pixels, showing that imagers can make efficient use of the far infrared arrays by using pixels that are a reasonably large portion of the Airy pattern. From other experiments, we conclude that this favorable result arises only if the telescope is substepped on a finer grid than the pixel-to-pixel spacing. In addition, the requirements on the accuracy of calibration and the knowledge of the point source response will be increased as the pixel size is increased. These factors must all be considered in the design of realistic systems for high sensitivity operation in the far infrared.

Although we have treated our simulations as if there is a hard confusion limit, at higher ratios of signal to noise it is likely that the source could be localized more precisely than in our experiments, leading to a smaller effective beam diameter and an improvement in limiting flux density. Again, achieving these goals will place greater demands on the calibration and knowledge of the point source response and therefore will depend on the details of system design.

As examples, we compute for an 85 cm telescope at 60, 100, and 150 μm the rms confusion noise limits due to distant galaxies given in Table I. The confusion noise from point sources with uncorrelated positions on the sky will scale in this case inversely as the effective aperture area, or as D^{-2}.

TABLE I

Noise Components for 1000 Sec Integrations

Noise Source	60 μm	100 μm	150 μm
Galaxies	$\sim$ 25μJy	350μJy	1500μJy
IR Cirrus	6μJy	89μJy	300μJy
Photon Noise	90μJy	140μJy	400μJy
Net Noise	94μJy	387μJy	1580μJy

3. Confusion by Infrared Cirrus

We base our discussion of confusion by infrared cirrus on the formalism of Gautier et al., (1992). To apply these results, we need to set three parameters: 1.) the power law index, α, of the cirrus emission power; 2.) the power at 0.01 arcmin^{-1}, $P_o(0.01/\text{arcmin})$; and 3.) the observing geometry.

Analysis of the IRAS far infrared data shows that the cirrus power spectrum has a spectral index of $\alpha = -2.6$ to -3.2 (Gautier et al., 1992, T. N. Gautier unpublished work). The IRAS data do not extend to the high spatial frequencies that are important for future missions that will have both telescopes of larger aperture and imaging far infrared arrays. However, the power spectrum can be estimated at these high frequencies by comparison with deep CCD exposures in the visible, which can reach 1 arcsec^{-1}. Cutri (private communication) finds that the power spectrum steepens slightly from the low spatial frequencies (i.e., α becomes more negative). In the following, we will assume $\alpha = -3.0$.

The level of cirrus contamination is characterized by P_o. As a medium level of cirrus contamination, we adopt the average P_o for $|b| > 50°$, namely 3×10^6 Jy^2/sr (Gautier et al., 1992). Other levels of cirrus emission are listed in Table II.

TABLE II

Cirrus Confusion Noise at 100 μm

Type of Area	$P_o(0.01\ \text{arcmin}^{-1})$ (Jy^2/sr)	RMS Confusion (μJy/beam)
Best 100 deg^2 in Baade's Hole	$\leq 1.2 \times 10^4$	≤ 6
500 additional deg^2 in Baade's Hole	$\leq 5 \times 10^5$	≤ 37
Average, $\lvert b \rvert \geq 50°$	3×10^6	89
Moderately Bright Cloud	10^8	510
Bright Cloud	10^{10}	5100

Because of the steep power spectrum of the cirrus emission, there is a premium in keeping the reference field as tightly coupled to the source as possible. We have therefore computed the cirrus confusion noise assuming an annular reference field that lies between 0.6 and 1.6 λ/D. For comparison with the galaxy confusion limit, we have interpolated to an effective beam diameter of 0.8 λ/D. Because of the low power at high spatial frequencies, the noise from cirrus confusion is only weakly dependent on the beam diameter but very strongly on the reference area size.

For the case of an 85 cm telescope, the resulting cirrus confusion noise limit at the average level of high latitude cirrus emission is 89 μJy per beam, with these assumptions. Values for the rms confusion noise for the 85 cm telescope at other wavelengths and in other parts of the sky are entered in Table II. The confusion noise will scale as the square root of P_o and as $D^{(\alpha/2-1)}$, i.e., as $D^{-2.5}$ for $\alpha = -3$.

4. Photon Noise

A far-infrared astronomical system will also be limited by photon noise. The cosmic background at infrared wavelengths can be broken down into a number of important components. At shorter wavelengths scattered solar radiation and thermal emission from zodiacal dust dominate, while beyond $\sim 40\ \mu$m emission from cold galactic dust must be included. Each of these components has a distinctive positional dependence on the sky, and large variations in the relative contributions are present. To investigate a case appropriate for extragalactic astronomy, we have taken the COBE Diffuse Infrared Background Experiment (DIRBE) measurements of the South Galactic Pole (Hauser et al., 1991). The noise from this background is computed in a conventional manner (e.g., Rieke, 1994); because the astronomical backgrounds are due to very dilute graybodies, the Boson correction can be ignored (van Vliet, 1967).

Since the focal plane fully samples the point source response, the signal is spread over a number of pixels. Therefore, we combine the measurements from the number of pixels needed to synthesize a point source, quadratically adding the noise. An additional noise penalty of a factor of 1.1 was added to allow for noise in the flat field. We have assumed a diffraction limited beam (for a telescope with 15% areal obscuration) and have computed the net signal in apertures centered on the source after subtraction of the average surface brightness in a reference area between 1 and 2.5 λ/D. The result is that the signal to noise for a point source is 8.2 times worse than that computed for a single 0.4 λ/D pixel (assuming all the signal fell on the single pixel) if the signal is measured in a 0.8 λ/D beam and 7.1 times worse if measured in a 1.2 λ/D beam. The latter beam size has the maximum ratio of signal to noise if photon noise alone is considered; however, to maximize the ratio of signal to noise that can be achieved in a confusion-limited situation, we will use the former beam size in the following.

Some degradation of the photon noise will occur in space because cosmic rays striking the detector will limit the integration times. A discussion of the extent of this effect is given by Herter (1990). The results depend on detector geometry, read noise, and the method of operation of the readout. We assume appropriate parameters for the far infrared arrays under development for SIRTF (Young et al., 1993), i.e., that a cosmic ray hit destroys all information in the integration after the hit, a hit affects the pixels on either side of the hit one so that the resulting net pixel area is 0.13 cm^2, and the read noise is 50 electrons; the result is that the pure background-limited photon noise will be degraded by a factor of ~ 1.5.

The photon noise limits in Table I are computed according to the description above and include the degradation by cosmic radiation. Of course, for a diffraction limited telescope, these detection limits scale as D^{-2}.

5. Combined Noise

The various sources of noise are combined in Table I, where the cirrus emission has been taken at the average for high latitude sky. Quadratic combination may give a misleading answer if the cirrus and extragalactic confusion are similar in magnitude; simulations are needed for this circumstance. However, over most of the high latitude sky the cirrus noise is relatively small and the results of this paper, where extragalactic and cirrus confusion have been considered separately, should be applicable.

The behavior across the far infrared band progresses rapidly toward increasing levels of confusion with increasing wavelength, both because of the increase in the density of galaxies of a given flux density and because of the increase in the diffraction limited beam size. At 60 μm, integrations greater than 10,000 seconds are required to approach serious levels of confusion, while 200 seconds and 100 seconds are required respectively at 100 and 150 μm. It is possible with an 85 cm telescope to reach limits about two orders of magnitude deeper than the data in IRAS bands 3 and 4.

We note that the scaling of the noise sources, for a given and roughly optimized method of source extraction, is D^{-2} for point source confusion noise and photon noise and $D^{-2.5}$ for cirrus noise. Thus, the relative importance of these differing noise sources will change only slowly with telescope aperture. Any telescope designed to operate at fundamental detection limits in the far infrared should be cold enough so that the photon noise does not increase significantly compared with the natural background.

6. Conclusion

The advent of far infrared detector arrays will allow qualitatively new approaches to sensitive astronomical detection at these wavelengths. By applying source extraction techniques similar to CLEAN or DAOPHOT, the effective beam size can be reduced substantially with a resulting reduction in confusion noise compared with predictions for instruments with a small number of detectors or for aperture photometers. The same benefits can be achieved for operating in regions where there is strong infrared cirrus. To operate in this mode, it is essential that far infrared telescopes have highly reproducible, preferably diffraction limited, beam profiles.

Table I shows as an example the sensitivity that can be achieved with a cryogenic 85 cm telescope using our CLEAN-like algorithm and operating at the average level of IR cirrus at high Galactic latitude. At 60 μm, the instrument will not become confusion limited unless integrations longer than 10,000 sec are attempted.

Because the three types of limiting noise scale with almost the same power of the telescope aperture, their relative contributions will not change significantly for telescopes of other sizes. As a result, the well-recognized benefits of operating meter-class telescopes at temperatures cold enough that they are natural background limited also apply to larger instruments. For example, although a telescope operating at < 5K at 160 - 200 μm will become confusion limited in about 60 seconds, the same telescope at 8K would require 1000 seconds to reach the same sensitivity limit; at 15K it would require 100,000 seconds. Therefore, designs for future far infrared telescopes should emphasize operating them at low temperatures.

Acknowledgements

We greatly appreciate the hospitality of the conference organizers and helpful comments from G. Helou and J. Condon. The work reported here was supported by NASA as part of the SIRTF project.

References

Condon, J. J.: 1974, *ApJ* **188**, 279

Franceschini, A., Toffolatti, L., Mazzei, P., Danese, L., De Zotti, G.: 1991, *A & A Suppl.* **89**, 285

Gautier, T. N., Boulanger, F., Perault, M., Puget, J. L.: 1992, *AJ* **103**, 1313

Hacking, P., Houck, J. R.: 1987, *ApJS* **63**, 311

Hauser, M. G., Kelsall, T., Moseley, S. H., Silverberg, R. F., Murdock. T., Toller, G., Spiesman, W., Weiland, J.: 1991, 'The First 3 Minutes' in S. S. Holt, C. L. Bennett, and V. Trimble, ed(s)., *AIP Conf. Proc. 222*, Am. Inst. of Phys.:New York, 161

Helou, G., Beichman, C. A.: 1991, *Proc. 29th Liege International Astrophysical Colloquium* **ESA SP-314**, 117

Herter, T.: 1990, *IR Phys.* **30**, 97

Rieke, G. H.: 1994, *Detection of Light*, Cambridge University Press

van Vliet, K. M.: 1967, *App. Opt.* **6**, 1145

Young, E. T., Scutero, M. J., Rieke, G. H., Haller, E. E., Beeman, J. W.: 1993, *SPIE* **1946**, 68

BACKGROUND LIMITED INFRARED AND SUBMILLIMETER INSTRUMENTS

J.-M. LAMARRE, F.-X. DÉSERT, T. KIRCHNER
Institut d'Astrophysique Spatiale
Bât 121, Université Paris XI,
91405 Orsay Cedex France

Abstract. The temperature and emissivity of infrared and submillimeter telescopes are basic parameters that drive the optical and thermal design of astronomical space projects. They determine also, among other parameters, the self-emission of the instrument and the photon noise produced by this radiation on the detectors. By comparing the telescope brightness with that of the sky in the 1μm–1 cm wavelength range, general conditions for background limited photometry are derived. For $\lambda < 0.4$ mm, temperature is the driving parameter, and for $\lambda > 0.4$ mm, temperature and emissivity have equivalent importances. It can be shown on actual projects that these two regimes determine different optical and thermal concepts. Although based on a simplistic approach, this work intends to help designers to handle some basic system parameters of infrared and submillimeter instruments.

Key words: Space instrumentation – Visible and Infrared diffuse backgrounds

1. Introduction

The fundamental limit to the sensitivity of observations is the photon noise of the detected radiation, i.e. the statistical variation with time of the number of photons that can be detected in a given measurement process. These intensity fluctuations are due to the quantum nature of photons and are part of the radiation itself. In laboratory experiments, it is possible to reduce their amplitude, at the expense of increased phase fluctuations. In the case of astronomical observations, they seem to be the ultimate limit to the precision of radiation measurements: the fluctuation of the part of the emitted radiation which is collected by a telescope is the smallest possible noise with this telescope. This ideal photometry (BLIP, or background limited photometry) can be met if the detector noise is smaller than the photon noise, which is possible now in infrared astronomy due to the significant advances that took place in the domain of detector technology in the one micrometer to one millimeter wavelength range. BLIP operation is therefore a commitment for new astronomical instruments unless other requirements, such as the presence of atmosphere for ground-based telescopes, or the difficulty of cooling large antennas for space projects, forbid it.

In the "phase space" of relevant parameters of space infrared to radio instruments, this paper addresses only the temperature and the emissivity of the instrument and from only one point of view, which is the photon noise. Other questions, such as angular or spectral resolution appear only as

Space Science Reviews **74**: 27–36, 1995.

external constraints. From this simplistic approach, we expect to reach conclusions general and consistent enough to be a guideline for the astronomers who design (or just dream of) future instruments.

In the next section, we study how photon noise is related to the power reaching the detector and derive a condition on the temperature and the emissivity of the instrument for BLIP operation for a given background. In section 3, the astronomical background is described, and the corresponding photon noise is derived. In section 4, we study the emissivity of instruments and in section 5 we derive a maximum instrument temperature for BLIP conditions, depending on wavelength and instrument emissivity. Section 6 is dedicated to comments on the position of projects under study with respect to the BLIP condition.

2. Photon noise

Two sources producing the same power on a detector do not necessary produce the same photon noise. The noise equivalent power produced by photon detection statistics in a given detection process with a thermal source is (Lamarre, 1986):

$$NEP_{\mathrm{ph}}^2 = \frac{2}{\eta^2}\int h\nu Q_\nu \, d\nu + \frac{(1+P^2)}{\eta^2}\int \Delta(\nu) Q_\nu^2 \, d\nu, \qquad (1)$$

where η is the quantum efficiency of the detector, Q_ν is the power reaching the detector per unit of optical frequency, P is the polarization degree of this radiation, and $\Delta(\nu)$ is its partial coherence factor.

$\Delta(\nu)$ is usually expressed, as a first approximation, as the inverse of the number of space modes of the instrument, i.e. $A\Omega/\lambda^2$, where $A\Omega$ is the beam throughput of the measurement process and λ the wavelength. This is one of the difficulties of this formula: the source producing the incident power may occupy only a small part of the instrument beam throughput. In other terms, it may produce coherent radiation ($\Delta(\nu) = 1$) in a multimoded instrument ($\Delta(\nu) \ll 1$). The opposite situation is often verified in real instruments. Nevertheless, if the beam throughput of the detector pixels is of the order of magnitude or smaller than λ^2, all types of sources will produce with these detectors nearly coherent detection processes. No significant difference in photon noise is to be expected for equal incident powers. A second source of difference is the possible polarization of the incoming radiation with a maximum consequence on the photon noise of a factor $\sqrt{2}$.

In consequence, for two different reasons, equal powers coming from the source and from the instrument itself may produce different photon noises. They are nevertheless expected to be small for imaging instruments and not very polarized sources. In the following part of this paper, it will be assumed that equal powers produce equal photon noise, while keeping in mind that this is only an approximation acceptable for the study of orders of magnitude.

Stating that the photon noise is dominated by the sky is then equivalent to writing the condition:

$$I_{\nu,\,\mathrm{instr}} = K I_{\nu,\,\mathrm{bg}}, \tag{2}$$

where $I_{\nu,\,\mathrm{instr}}$ and $I_{\nu,\,\mathrm{bg}}$ represent the specific "background" intensity produced respectively by the instrument itself and by the source on the detector, and K is an arbitrary constant smaller than one. When $K = 1$, the background power cannot be lowered by more than a factor of two, which, at best yields a factor of two on the photon noise. Spending energy to still reduce that noise may not be worth.

If we consider the instrument as a greybody of emissivity $\epsilon(\nu)$ and of temperature T, it produces a specific intensity

$$I_{\nu,\,\mathrm{instr}} = \frac{2\epsilon h\nu^3}{c^2} \frac{1}{e^{h\nu/kT} - 1}. \tag{3}$$

Then, the condition expressed by equation (2) becomes:

$$T = \frac{h\nu}{k} \frac{1}{\ln(1 + \frac{2\epsilon h\nu^3}{c^2 K I_{\nu,\,\mathrm{bg}}})}, \tag{4}$$

which defines the maximum temperature that keeps the photon flux produced by the instrument itself (in the absence of other sources) to a value equal to K times the sky background. This equation is based on an approximation. It does not take into account the intrinsic noise of the detectors or any other limitations such as confusion. It has nevertheless the advantage of simplicity. It gives the occasion to handle two basic parameters of instrumentation (temperature and emissivity) with a very clear criterion that does not depend on angular resolution, spectral resolution, detection principles, number of pixels, and so on.

3. The astronomical background

Although space experiments get rid of the strong atmospheric disturbances (opacity and emission), they still face faint but present foregrounds and backgrounds (Figure 1a). Interplanetary dust which pervades the inner Solar System scatters the Sun light in the near infrared and emits a thermal spectrum throughout the mid infrared as observed by IRAS (Hauser et al, 1984). Faint stars make a near infrared galactic background (Puget, 1976) and interstellar dust emission throughout the infrared (Boulanger & Pérault, 1988) produces a highly structured background (the so-called cirrus clouds) even at high galactic latitudes and which probably contains mid-infrared spectral features (Sellgren et al., 1985). The submillimeter background is due to the 3K cosmic background radiation.

The photon noise produced by these backgrounds is not negligible. Let us consider a perfect instrument with a transmission and a quantum efficiency equal to one, with a beam throughput equal to that of the diffraction limit, and with spectral resolutions of resp. 4, 10^2, and 10^4. As shown in Figure 1b, the corresponding NEPs are in the range of 10^{-17} to 10^{-20} W $\mathrm{Hz}^{-1/2}$, which is reached by the most recent IR and submillimeter detectors. Therefore, the question of BLIP operation for astronomical space IR instruments is not a purely academic one. It is a requirement or a design goal that must be, and that is effectively at the center of the design of these instruments.

4. Instrument emissivities

4.1. Telescope designs

Is it necessary to design low emissivity IR instruments? When looking at current IR and submillimeter projects, it is clear that different answers have been given to this question. Since it is the largest part in instrument design, the telescope is often also its warmest optical part. The focal optical system is usually cooled to temperatures low enough to give insignificant contributions to the general backgrounds. Therefore, the designs differ mainly by the telescope. In order to quantify the impact of instrument emissivities on their required temperature, we have chosen three different designs with supposed emissivities that cover the range of realistic ones.

a) High emissivity instrument based on a Cassegrain telescope. The baffles have been designed in order to minimize straylight, which may lead to significant obscurations by emissive elements. SIRTF (Werner & Simmons, 1994) may be an example of this design. We suppose that in this case the telescope has an emissivity of 0.1 plus that of the two mirrors:

$$\epsilon_{\mathrm{tel}} = 0.1 + 2\epsilon_{\mathrm{m}}, \tag{5}$$

where ϵ_{m} is the mirror emissivity.

b) Low emissivity Cassegrain telescope. For mechanical or size reasons, a Cassegrain design was chosen, but in the same time, low emissivity was a design goal, and emissive items have been reduced to the strict minimum, such as the legs of the spider, and/or part of the obstruction by the secondary mirror. FIRST (Beckwidth et al., 1993) may be an example of this design. The emissivity of this type of telescope may be given by:

$$\epsilon_{\mathrm{tel}} = 0.03 + 2\epsilon_{\mathrm{m}}. \tag{6}$$

c) Low emissivity off-axis telescopes. An example of this type of telscope is the tilted off-axis Gregory design of the SAMBA or FIRE projects (e.g.

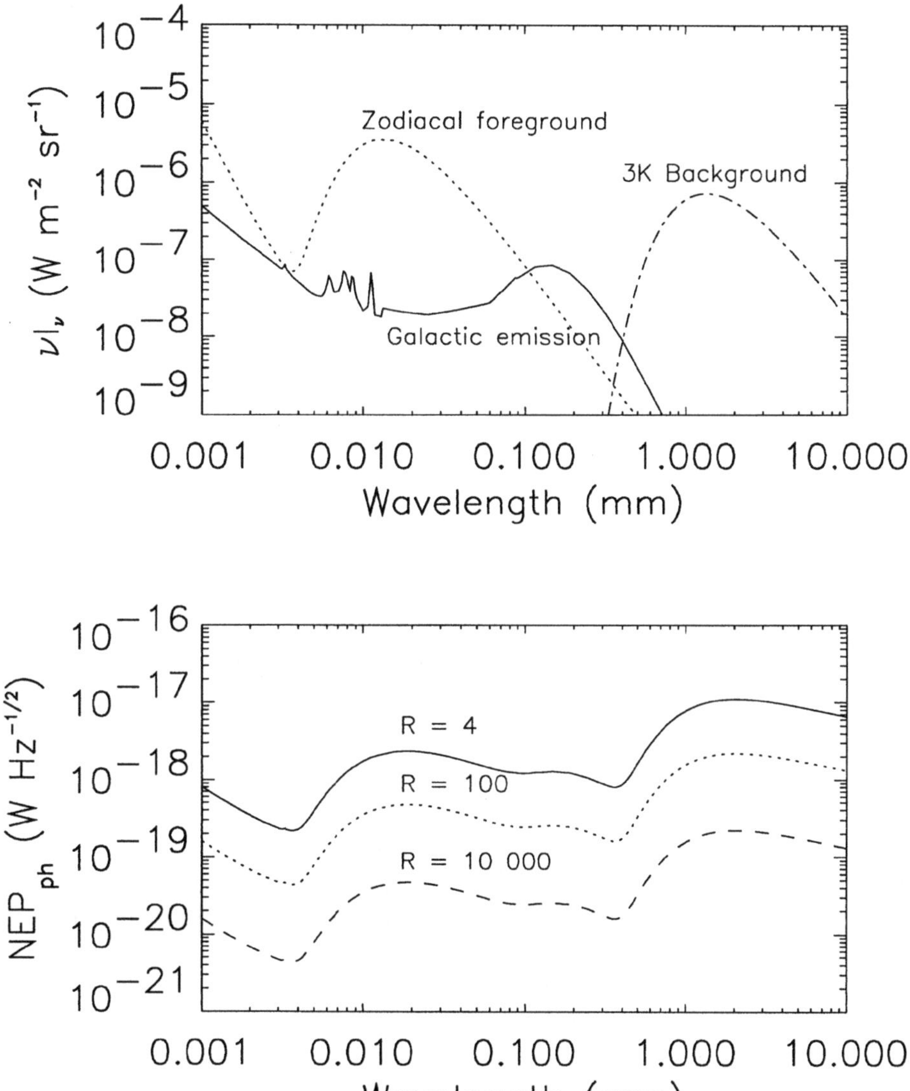

Fig. 1. a) Various foregrounds and backgrounds that space experiments are facing. The zodiacal foreground is a lower limit obtained only when observing near the ecliptic poles. b) Noise Equivalent Power as a function of wavelength due to the astronomical background (shown in Figure 1a) outside the atmosphere for a perfect instrument at the diffraction limit with a resolution of resp. 4, 10^2, and 10^4.

Lange et al., 1994). There is no obstruction, and the design of the focal optics includes a cold Lyot stop that reduces the instrument signal to that of the oversized telescope mirrors.

$$\epsilon_{\mathrm{tel}} = 2\epsilon_{\mathrm{m}}. \tag{7}$$

4.2. Mirror emissivities

The reflectivity R of bulk polished metal is given, following the theory of Hagen & Rubens, by Born & Wolf (1990):

$$R_{\mathrm{m}} \approx 1 - 2\sqrt{\frac{\nu}{\sigma}} + \dots \tag{8}$$

where ν is the optical frequency and σ the DC current resistivity of the metal. Following Kirchhoff's law, for bodies in local thermal equilibrium, emissivity is equal to absorptivity, and in this case, to $1 - R_{\mathrm{m}}$.

$$\epsilon_{\mathrm{m}} \approx 2\sqrt{\frac{\nu}{\sigma}} + \dots \tag{9}$$

Nevertheless, several factors such as imperfect layers of metal, contamination, and micro-roughness can increase the emissivity of actual mirrors. For the LDR studies, P. Swanson (1982) found that a large number of measurements of different metallic samples could be fitted by a common law:

$$\epsilon_m = 0.1\lambda_{\mu\mathrm{m}}^{-1/2}. \tag{10}$$

This formula is based on reflectivity measurements of gold, silver, and aluminum made by several authors (Touloukian & Dewitt, 1970, Weiss, 1980, and Otoshi & Thom, 1981) between a few microns and several centimeters of wavelength. The equal figures found for different metals seem to demonstrate that phenomena other than pure resistivity are dominant. The emissivity is simply taken, by supposing that scattering is negligible, as equal to one minus the reflectivity, which maximizes ϵ. The loss of resistivity of metals at low temperature is not taken into account, in spite of the fact that telescopes of IR and Submm space projects are usually cooled by cryogenic fluids or passively. The derived emissivities are several times larger than the theoretical value of equation (9) or than the best experimental results (Toscano & Cravalho, 1976, and Padalka & Shklyarevskii, 1961). This estimation is therefore pessimistic for freshly made clean mirrors, but may be a realistic approach for actual space projects submitted to different types of contamination all along their life. P. Swanson's law will be used in the following sections of this paper.

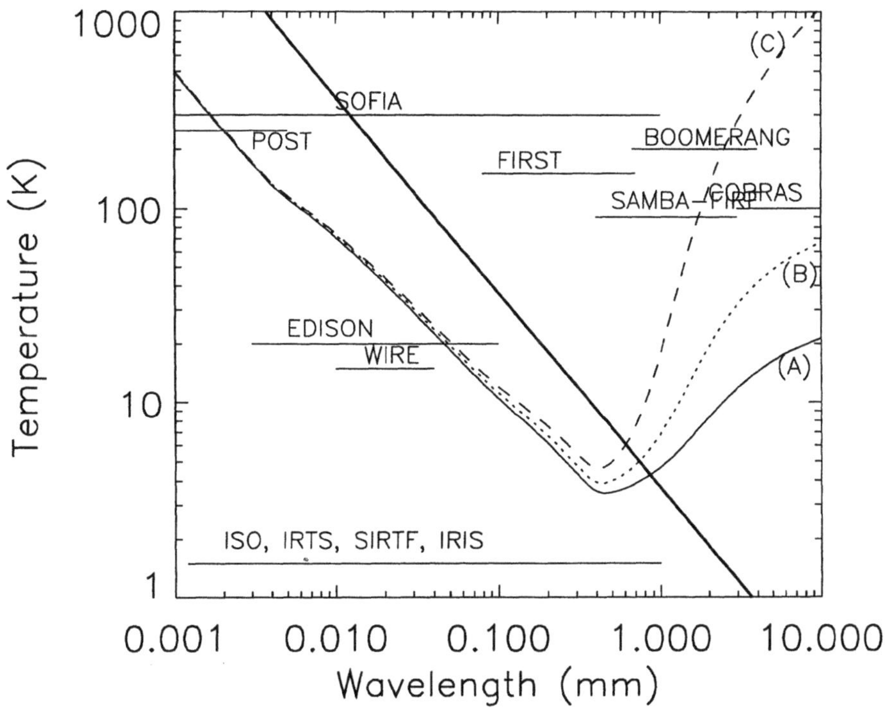

Fig. 2. Maximum allowed instrument temperature for the experiment to be background limited. A rough working range of wavelengths and temperature for a series of planned experiments are also indicated, see this conference proceedings for more details. Lines A, B, and C refer to Section 4.1 concerning the telescope configuration (equations 5, 6, and 7 resp.). The thick line represents Wien's displacement law for the maximum of νI_ν of a blackbody

5. Results

Figure 2 is obtained by applying equation (4) with $K = 1$ to the three cases defined in Section 4.1. The curves show the temperatures of the three types of telescope that produce backgrounds equal to that of the sky. These curves present a pronounced minimum around $400\,\mu$m. The features of the sky background cannot be easily recognized. The general shape of the curves is driven by Planck's function (PF) that varies much more than the sky background. This fact determines two regimes corresponding to the well known limit cases of the blackbody radiation, the Wien and the Rayleigh-Jeans (R–J) regimes.

The left part of the curve $\lambda \leq 400\,\mu$m corresponds to the short wavelength cut–on of the PF, which is very steep. Due to Wien's displacement

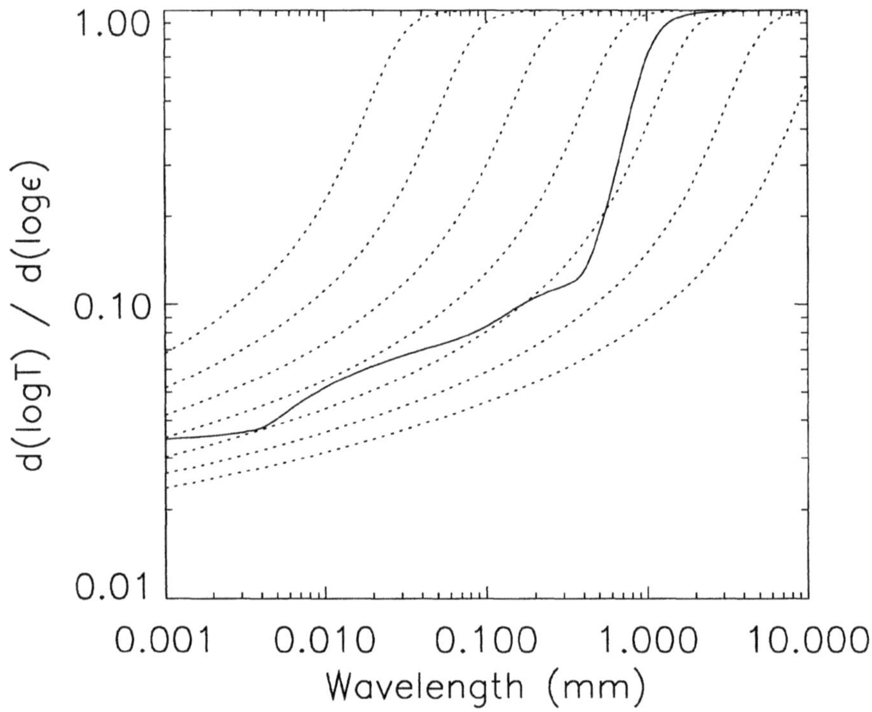

Fig. 3. The relative influence of temperature and emissivity in having a background limited instrument. The parameter R (see eqn. 11) is shown as a function of wavelength for a background going from $10\,\mathrm{Wm^{-2}sr^{-1}}$ (upper dashed curve) to $10^{-11}\,\mathrm{Wm^{-2}sr^{-1}}$ (lower dashed curve) by powers of 100. The solid line corresponds to the sky background met by space experiments (see Figure 1a). All the curves were evaluated for a C configuration (see Section 4.1).

law, at each temperature corresponds a cut–on wavelength that does not change significantly with the telescope emissivity. The longwave part of the curve corresponds to the smoother R–J regime, where the emissivity and temperature have identical effects and where the three curves show diffferent behaviors. The relative influences of emissivity and temperature on the instrument self–emission can be represented on Figure 3, where the ratio of the derivatives of the self–emission with respect to the emissivity and the temperature is plotted:

$$R(\lambda, \nu I_{\nu,\mathrm{instr}}) \equiv \frac{\partial \ln(\nu I_{\nu,\mathrm{instr}})/\partial \ln \epsilon}{\partial \ln(\nu I_{\nu,\mathrm{instr}})/\partial \ln T} = \frac{y}{(1+y)\ln(1+y)}, \tag{11}$$

where $y = 2\epsilon h\nu^3/(c^2 K I_{\nu,\,\mathrm{instr}})$. The dotted lines represent R for values of $(\nu I_{\nu,\,\mathrm{instr}})$ taking constant values from 10^{-11} to $10\,\mathrm{Wm^{-2}sr^{-1}}$. They show that $R \approx 1$ at long wavelengths and $R < 0.1$ at short ones. The solid line represents the value of R for the sky background of Figure 1a.

Figures 2 and 3 help to define the two regimes concerning the efficiency of temperature and emissivity to meet the BLIP condition:

a) For $\lambda \leq 400\,\mu\mathrm{m}$ the temperature is the dominant parameter. The needed temperature is nearly proportional to the radiation frequency and independent of emissivity. This demonstrates that in this domain, reducing the emissivity is not and efficient way of meeting the BLIP condition. Very low temperatures are mandatory and directly related to the maximum wavelength of the project.

b) For $\lambda \geq 400\,\mu\mathrm{m}$, the emissivity and temperature parameters have effects of the same order of magnitude. In this regime, one must put in balance the reduction of temperature and emissivity. Low emissivity designs must be used.

These two types of experiments have been designed and presented during this conference. Figure 2 represents, together with the BLIP limit curves, the temperature and spectral coverage of many of these experiments. The data used for this diagram may be subject to some errors due to the changing nature of projects and to the use of oral information (see this proceedings for more precise values). Nevertheless, it is possible to distinguish three families of experiments, two of which corresponding to the two regimes of the BLIP condition, and the third one not being background limited.

The "shortward BLIP" instruments include POST, EDISON, WIRE, ISO, IRIS, SIRTF and IRTS. They are usually not designed for a low emissivity. The warmest experiments (POST and EDISON) are not background limited in their whole spectral range. In this case, it will not be possible to meet the BLIP condition by decreasing their emissivity. Nevertheless, if it is not possible to cool them to lower temperatures, a low emissivity design will reduce the photon noise and increase the sensitivity. These instruments compete with warm ones, and may be more efficient for given modes of an observation.

The "longward BLIP" instruments are COBRAS, BOOMERANG, SAMBA and FIRE. All of them have low emissivity designs using off–axis tilted gregorian telescopes. SAMBA–FIRE and BOOMERANG do not meet the BLIP condition in the complete spectral range, since this condition would necessitate to cool the balloon–borne BOOMERANG experiment down to 30K and the SAMBA–FIRE telescope to 2K, changing the cost class of these projects.

The "non–BLIP" instruments are the airborne SOFIA telescope and the space FIRST experiment. The SOFIA telescope takes the temperature of the atmosphere at the flight altitude and it would not be possible to build a cryostat around a 3 meter class FIRST telescope. A low emissivity design would reduce the photon noise in these two projects but would be more difficult to implement in the case of the airborne telescope.

6. Conclusions

The analysis of the conditions for Background limited photometry of astronomical instruments by using only emissivity and temperature as parameters proved to successfully describe the situation of infrared and submillimeter instruments with respect to these two parameters. Three classes of instruments have been found, two of which corresponding to the two regimes of background limited photometry. General information that could be used to improve the instrument design were derived from this study that does not take into account confusion problems, detector type, and spectral resolution.

References

Beckwith, S., Cornelisse, J., Van Dishoeck, E., Encrenaz, P., Genzel, R., Griffin, M., Harris, A., Hills, R., de Jong, T., Kollberg, E., Kreysa, E., Lamarre, J.-M., Lellouch, E., Martin-Pintado, J., Natale, V., Pilbratt, G., Poglitsch, A., Puget, J.-L., Rowan-Robinson, M., Stutski, J., Volonté, S., Whyborn, N.: 1993, 'FIRST, Far Infrared and Submillimetre Space Telescope', *ESA SCI(93)6.*

Born, M., Wolf, E.: 1990, *Principles of Optics, Sixth (corrected) edition*, Pergamon Press, 622

Boulanger, F., Pérault, M.: 1988, *ApJ* **330**, 848

Hauser, M. G., et al.: 1984, *ApJLett* **285**, L31

Lamarre, J.-M.: 1986, 'Photon noise in photometric instruments at far-infrared and submillimeter wavelengths', *Appl. Opt.* **25**, 870-876

Lange, A. E., Bock, J. J., Mason, P.: 1994, *this conference.*

Otoshi, T. Y., Thom, E. H.: 1981, 'Surface Resistivity measurements of candidate subreflecto surfaces', *TDA Progress Report*, **42-65**, pp. 142-150, JPL.

Padalka, V. G., Shklyarevskii, I. N.: 1961, *Optics and Spectrocopy* **11**, 285

Puget, J.-L.: 1976, *COBE internal report.*

Sellgren, K., et al.: 1985, *ApJ* **299**, 416

Swanson, P.: LDR studies in 1982, private communication.

Toscano, W. M., Cravalho, E. G.: 1976, *Trans. ASME* **C98**, 438

Touloukian, Y. S., Dewitt, D. P.: 1970, 'Thermal radiative properties, metallic elements and alloys', *Thermophysical properties of matter*, **vol. 7**, Plenum, New York.

Weiss, R.: 1980, 'Measurements of the cosmic background radiation', *ARA&A* **18**, 489-535

Werner, M. W., Simmons, L. J.: 1994, *this conference.*

SIMULATIONS OF THE MICROWAVE SKY AND OF ITS "OBSERVATIONS"

F.R. BOUCHET and R. GISPERT
Institut d'Astrophysique, Paris, and Institut d'Astrophysique Spatiale, Orsay, France

N. AGHANIM
Institut d'Astrophysique Spatiale, Orsay, France

J.R. BOND
CITA, Toronto, Canada

A. DE LUCA
Institut d'Astrophysique Spatiale, Orsay, France

E. HIVON
Institut d'Astrophysique, Paris, France

and

B. MAFFEI
Institut d'Astrophysique Spatiale, Orsay, France

Abstract.
Here follows a preliminary report on the construction of fake millimeter and sub-millimeter skies, as observed by virtual instruments, *e.g.* the COBRA/SAMBA mission, using theoretical modeling and data extrapolations. Our goal is to create maps as realistic as possible of the relevant physical contributions which may contribute to the detected signals. This astrophysical modeling is followed by simulations of the measurement process itself by a given instrumental configuration. This will enable a precise determination of what can and cannot be achieved with a particular experimental configuration, and provide a feedback on how to improve the overall design. It is a key step on the way to define procedures for the separation of the different physical processes in the future observed maps. Note that this tool will also prove useful in preparing and analyzing current (*e.g.* balloon borne) Microwave Background experiments.

Key words: Cosmology – Microwave Background Anisotropies.

1. Introduction

The primary objective of the COBRA/SAMBA mission submitted to ESA and of the SAMBA mission submitted to CNES is to determine the spatial characteristics of the fluctuations of the Cosmological Microwave Background radiation (hereafter CMB), at all angular scales between five minutes of arc and a few degrees. In order to reach that goal with the required precision (detection of signals with an equivalent $\Delta T/T \simeq 10^{-6}$, *i.e.* $\simeq$ 3 μK), it is necessary to separate the numerous possible contributions to this radiation. We have embarked on the construction of simulated millimeter and sub-millimeter skies, using theoretical modeling and data extrapolations, which we then observe with virtual instruments, *e.g.* those on the COBRA/SAMBA payload. Our goal is to create maps as realistic as possible

Space Science Reviews **74**: 37–43, 1995.

of the relevant physical phenomena which may contribute to the detected signals.

The usual approach of comparing the expected variances of the distribution of fluctuations of different types (*i.e.* their *rms* values) is not sufficient because these distribution functions are certainly non-Gaussian for some of the sources. For example, the submillimeter emission of resolved galaxies and the Sunyaev–Zeldovich effect due to clusters are certainly in that category since there are localized objects. But this might also be true for other sources of fluctuations such as those arising from topological defects, remnants of an earlier phase in the early Universe. As another example, the emission of the dust itself at the maximum wavelength of the CBR, either from the Milky Way or from other galaxies, can probably be removed by spectral extrapolations from maps at shorter wavelengths. In that case, we wish to study the spatial properties of the residuals under various assumptions concerning, *e.g.* the respective distribution of a cold and hot dust component. Another reason is that the resulting maps may be used as test beds for ideas concerning the data analysis process. We hope to check then the feasibility of various signal-separation techniques. Furthermore, we plan to use this tool in an iterative fashion in order to optimize the instrumental characteristics of the planned experiment in order to best discriminate both spatially and spectrally the various components.

2. Sky Simulations

The sources of flux anisotropies which may contribute at the wavelengths of the instrument (say between $200\,\mu$m and 2 cm) can be decomposed as follows:

- Primary $\Delta T/T$ of the 2.726 K background. Those are imprinted predominantly during the last Thomson scatterings of photons by free electrons.
- Sunyaev-Zeldovich (SZ) effect from the hot gas in galaxy clusters. It may in turn be decomposed into 2 pieces:
  - y-distortions (thermal part), due to the scattering of "cold" CMB photons off the "hot" electron of the ionized intracluster gas
  - $\Delta T/T$ distortions (kinematical part), due to the Doppler shift from the clusters peculiar velocities accompanying Thomson scattering
- Submillimeter emission from the galaxies, which can also be decomposed into two pieces:
  - contribution from spatially resolved galaxies
  - background fluctuations due to the integrated emission of unresolved galaxies (including starburst galaxies and AGN's)

- Emission from our own Galaxy. Stellar sources have a weak contribution at the wavelength of interest, but there are three main other components which are likely to be relevant:
  - Interstellar dust (even at high galactic latitude: galactic cirrus)
  - Bremsstrahlung emission
  - Synchrotron radiation
- Other sources of fluctuations. Let us mention just a few: topological defects created during an early Universe Symmetry–breaking phase transition (*e.g.* Cosmic Strings, Textures, Global monopoles, etc...), or, in the Earth neighborhood, Zodiacal light or asteroids trails.

At this preliminary stage, our modeling of the various physical phenomena has been done as follows:

- For the primary $\Delta T/T$ and the SZ effect from clusters (both y & $\Delta T/T$) we have created realizations corresponding to specific theoretical models (*e.g.* standard CDM). This was also done for the possible secondary fluctuations from cosmic strings.
- For spatially resolved galaxies and the galactic dust, as a simple first step, we have split $100\,\mu$m ISSA-IRAS maps into two maps. The first one is a population of resolved extragalactic sources assumed to be at a uniform temperature $T = 30$ K (with a dust standard emissivity $\varepsilon_\nu \propto \nu^{-2}$). The second one is the dust galactic background which we assumed to be also at a uniform temperature, $T = 18$ K.
- For the synchrotron radiation, we have simply extrapolated the 408 MHz maps from Haslam (1982) (using a $I_\nu \propto \nu^{-0.7}$ law).

All these processes can be stored as elementary maps (presently $12.5° \times 12.5°$, with 500^2 pixels, each of size 1.5 arcminutes) together with conversion coefficients to energy flux density at a given frequency.

We are currently modeling the fluctuations from unresolved galaxies by Monte-Carlo simulations. We draw a population of galaxies by redshift interval. In each interval, we take into account the spatial correlations between objects and their luminosity functions, according to their (assumed) time–evolution. The spectra of the galaxies are determined from their luminosity and type. The results of this study are not included yet. We plan to use more precise models of dust emissivity as in Désert et al. (1990). And the 60μm IRAS maps and DIRBE-COBE data at high latitude will furthermore permit us to actually deduce the dust temperature fluctuations. We also urgently need to include estimates of the Bremsstrahlung emission (for the COBRA wavelength), possibly using COBE maps or Hα maps extrapolated to the small angular scales of relevance to us.

3. Satellite Measurements Simulations

All of our virtual instruments have at this stage the following characteristics:

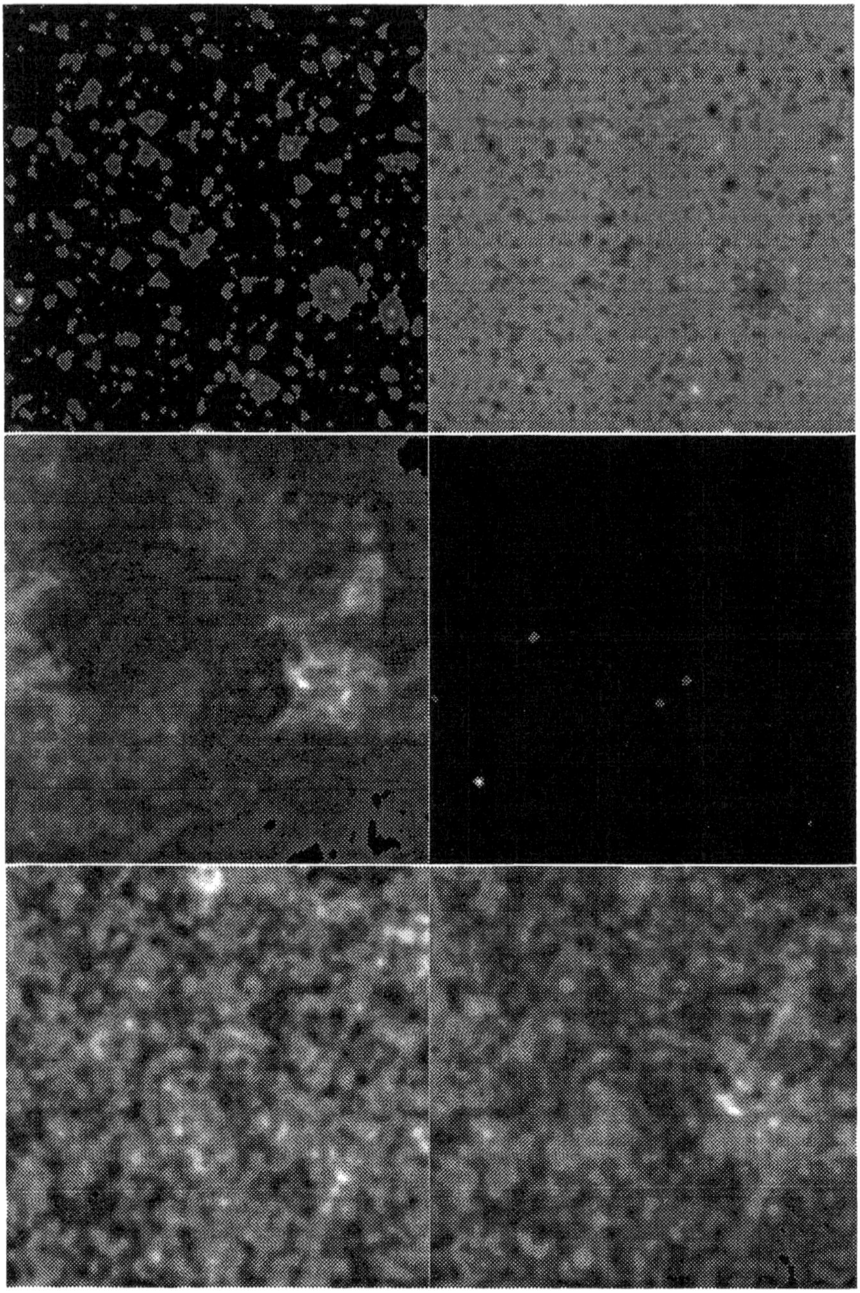

Fig. 1. $10° \times 10°$ maps in the 1.44 mm band for a gaussian lobe of 7.5' FWHM. They correspond, from left to right, and from top to bottom, to the thermal and kinematical SZ effect, to the dust and foreground galaxies, to CDM primary CMB fluctuations and the sum of all the above plus the (not shown) synchrotron contribution.

- nominal spectral bands with unity transmission across the entire band
- Gaussian lobes of FWHM equal to the diffraction limit
- a $1/f$ detector noise with a low frequency cutoff
- the orbital type is a simplified model assuming full sky coverage, but with integration time varying from 30s to 1000s per pixel.

The maps will also depend on the accuracy level retained for the digital data sent to the ground. One further issue will be the type, if any, of on-board compressing and pre-processing of the data. The structure of our tool is designed to deal easily with further modifications of the experimental configuration. Note that to avoid spurious effects on the boundaries, we keep only the inner $10° \times 10°$ of the maps, once convolved with a lobe.

4. Sample Results and Perspectives

Figure 1 shows a possible choice of 5 contributing processes and their resulting sum plus a (not shown, but included) synchrotron component. The scales of the different maps are independant. They correspond to a same band with uniform transmission of unity between 1.1 and 1.8 mm, and an observation with a gaussian beam of 7.5 minutes of arc. At this wavelength, the dominating fluctuations are those of the CMB. Still, it is clear that the pattern of the sum map is rather different from that of the CMB alone. See Table I for the contributions of the processes to the total variance in the other wavebands.

TABLE I

Contribution (in %) of each process to the total variance of the maps in each SAMBA channel. Note though that this is quite deceiving for non-gaussian contributions.

Mean λ(mm) (& $\Delta\lambda/\lambda$)	2.23 (0.4)	1.44 (0.5)	0.85 (0.7)	0.46 (0.6)
FWHM (arc-min)	10.5	7.5	4.5	3.0
Primary $\Delta T/T$	93.174	53.748	0.839	$\leq$0.001
SZ (y)	0.083	$\leq$0.001	0.006	$\leq$0.001
SZ ($\Delta T/T$)	0.002	0.002	$\leq$0.001	$\leq$0.001
Foreground galaxies	$\leq$0.001	$\leq$0.001	0.007	0.026
Galactic dust	6.737	46.249	99.148	99.973
Synchrotron	0.003	$\leq$0.001	$\leq$0.001	$\leq$0.001

We already have available analyses like distributions of pixel intensities, or C_ℓ decompositions (*i.e.* the coefficients of a standard spherical harmonics decomposition), see Figure 2. We are implementing measurements of the

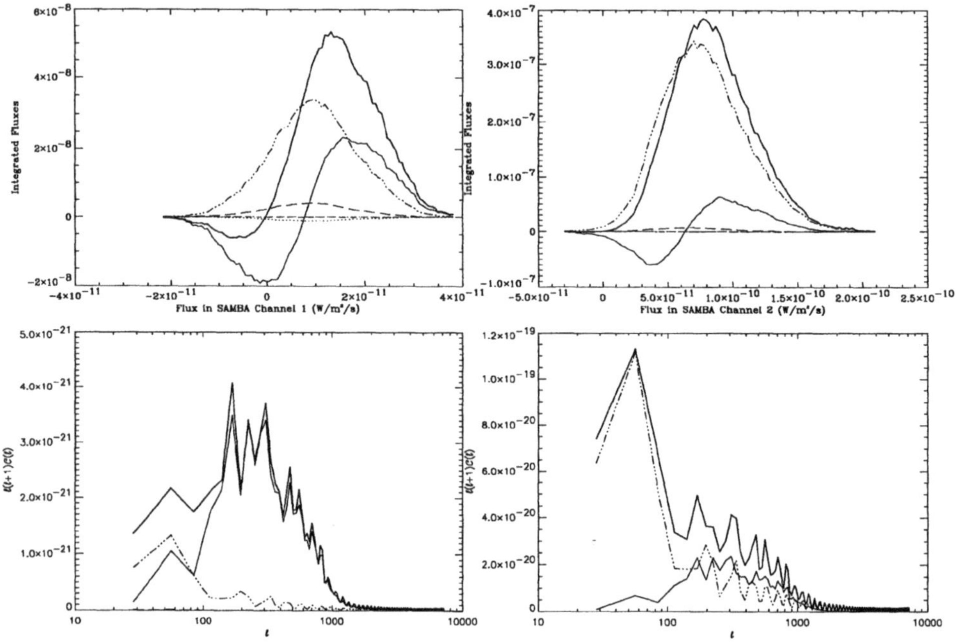

Fig. 2. The top panel shows the flux integrated over all pixels as a function of the total flux in a pixel, for the channels 1 (left) and 2 (right). The total, *i.e.* the x-axis value times the number of pixels with that flux, is given by the thick line. The CMB, $y - SZ$, galaxies, synchrotron, and dust contributions are denoted by a solid, dotted, dashed, dashed–triple–dotted line (the other components being undistinguishable from 0 on this plot). The corresponding angular contributions $\ell(\ell+1)C(\ell)$ are plotted below. The channels 3 & 4 are entirely dust–dominated.

angular correlation function of the anisotropies, $C(\theta)$, and of higher order statistics. We shall soon start investigating effective signal–separation techniques.

Of course, our current modeling is far from perfect. For the high–z sources of fluctuations, we have to rely on theoretical models. In some cases, years of effort have led to the development of reliable tools, such as for the primary $\Delta T/T$ fluctuations in Gaussian models. In other case, it is technically hard to make definite predictions, even though the model is perfectly well specified (*e.g.* for strings, or the SZ effect which depends on the gas behavior in very dense environments). We also have only weak constraints on the high–z distribution of galaxies. For lower–z sources of fluctuations, like those coming from the dust in galaxies, we cannot do much more at this time than to extrapolate the dust emission from the IRAS measurements. But there might very well be another cold dust population, spatially uncorrelated with the hot one. . . In any case, the database structure adopted should permit us to incorporate all the latest developments in relevant theoretical modeling and/or new observations, as they appear.

Acknowledgements

We thank F.-X. Désert, M. Giard, F. Pajot, M. Pérault, and J.-L. Puget for fruitful discussions and help in the launching of this project.

References

Bouchet, F.R., Bennett, D.P. and Stebbins, A.: 1988, *Nature* **335**, 410.
Désert, F.X., Boulanger, F., Puget, J.L.: 1990, *A&A* **237**, 215.
Maffeï B.: 1994, *PhD Thesis, to be published.*
Bond, J.R. and Myers, S.: 1994, *ApJ*, in press

RADIATIVE AND HYBRID COOLING OF INFRARED SPACE TELESCOPES

...and an example: The ≈ 5K "Very Cold Telescope" option for Edison

TIMOTHY G. HAWARDEN
Joint Astronomy Centre, 660 North A'ohoku Place, University Park, Hilo, Hawaii 96720, USA

ROBERT CRANE* and HARLEY A. THRONSON, JR.
Dept. of Engineering and Dept. of Physics and Astronomy, University of Wyoming, P O Box 3905 Laramie, Wyoming 82071, USA*

and

ALAN J. PENNY, ANNA H. ORLOWSKA and THOMAS W. BRADSHAW
Daresbury Rutherford Appleton Laboratory, Chilton, Didcot, OX11 0QX, UK

Abstract. The designs of cold space telescopes, cryogenic and radiatively cooled, are similar in most elements and both benefit from orbits distant from the Earth. In particular such orbits allow the anti-sunward side of radiatively-cooled spacecraft to be used to provide large cooling radiators for the individual radiation shields. Designs incorporating these features have predicted T_{tel} near 20 K. The attainability of such temperatures is supported by limited practical experience (IRAS, COBE). Supplementary cooling systems (cryogens, mechanical coolers) can be advantageously combined with radiative cooling in hybrid designs to provide robustness against deterioration and yet lower temperatures for detectors, instruments, and even the whole telescope. The possibility of such major additional gains is illustrated by the Very Cold Telescope option under study for *Edison*, which should offer $T_{tel} \leq 5$ K for a little extra mechanical cooling capacity.

Key words: Space Telescopes – Infrared – Radiative Cooling – Hybrid cooling

1. Introduction

Radiative cooling has of course been both a factor and a tool in spacecraft design since the earliest orbital missions. It has been widely exploited for cooling spacecraft subsystems, especially when infrared sensors are carried, *e.g.* for Earth surveillance. For infrared astronomy, the whole telescope needs to be cooled, in order to reduce the thermal radiation emitted by the optics, which otherwise generates a high background signal on the detectors. However experience with dedicated astronomical payloads is limited; indeed no dedicated astronomical mission which is mainly *radiatively* cooled has yet flown, so techniques have so far had to be inferred or transferred from other designs.

* Now at: Ford Motor Company, Lansing, Mich. USA

Space Science Reviews **74**: 45–56, 1995.

An extensive outline of basic design principles for the optimised use of radiative cooling for an entire payload was presented at the Edinburgh workshop on the Next Generation Infrared Space Observatory (Hawarden et al., 1992). The present paper outlines subsequent developments, in particular those associated with the combination of radiative cooling with other cooling methods, which appears to offer the potential both of very low temperatures and designs which are robust against deterioration and component failures during a long mission.

2. Designing for mainly-radiative cooling

2.1. Basic design of cold space telescopes

The basic design goals of all cold space telescopes are similar: to reduce as much as possible all heat flows from the outside into the cold telescope systems, as well as parasitic heating from, *e.g.*, detector electronics and mechanisms. This is done by intercepting and diverting as much as possible of the sun's heating with sunshades, in current designs usually double; and by protecting the cold systems from radiated heat, by surrounding them with multiple, preferably cooled, radiation shields, and from conducted heat, by supporting them with straps or struts with a high ratio of strength to conductivity.

Cryogenic missions use latent heat of vaporisation and the enthalpy of the boiloff gas to remove the remaining heat and to cool the radiation shields, while radiatively-cooled missions use radiation to space for both purposes: from the telescope tube to cool the optical system, from individual radiators to cool each of the radiation shields. Since the radiators are localised, this implies good internal conductivities in the shields and down the telescope tube.

Hawarden et al. (1992) emphasised the advantages for all types of cold space telescopes, but especially for radiatively cooled designs, of an orbit far from Earth, preferably non-geocentric. In addition to a lower overall heat load in the absence of Earth radiation, for mainly radiatively-cooled missions these include the reduction or elimination of aperture heating, which is more serious when the telescope tube must have high longitudinal conductance, and the freedom to provide each radiation shield with its own large and efficient cooling radiator on the rearward (antisunward) side of the spacecraft. The larger radiator areas produce important performance gains (in terms of reductions in $T_{\rm tel}$) relative to designs for geocentric orbits, which generally confine the radiators to the annuli separating the individual radiation shields.

The rewards of success are, for cryogenic missions, a longer lifetime with a given cryogen load, and for non-cryogenic missions, lower telescope and instrument temperatures.

2.2. Advantages and disadvantages of mainly-radiative cooling.

The potential advantages of a serious exploitation of radiative cooling as opposed to open-loop cryogenic cooling are now widely recognised and a number of proposals aim to exploit them. They are:

(i) The option to launch the telescope at room temperature and then allow it to cool down on orbit.

(ii) The absence of bulky cryogen tanks, enabling the telescope to occupy a much larger fraction of the available volume of the launcher fairing, permitting inclusion of a larger diameter telescope.

(iii) Substantial mass savings, arising from the elimination of the stout outer vacuum vessel, cryogen tanks and plumbing required by a mainly-cryogenic mission, again freeing resources to be directed towards a larger telescope.

(iv) The possibility of a much longer lifetime, not limited by cryogen supplies.

Amongst proposals described in these proceedings, the mainly-radiative *Edison* (Thronson et al., 1994) and FIRST (ESA's Far Infrared and Submillimetre Telescope; Cornelisse & Batchelor: 1993) plan to exploit all of these; the radiative/cryogenically cooled Space Infrared Telescope Facility (SIRTF; Werner & Simmons, 1994) exploits (i), (ii) and (iii) while the InfraRed Imaging Surveyor IRIS (Matsumoto, 1994), a fully hybrid mission using radiation, mechanical cooling and cryogens, to be launched cold, exploits (ii).

Since nothing comes for free, the attractions of radiative cooling must be balanced against certain drawbacks.

(i) If launched warm, a radiatively-cooled mission will require an appreciable time on orbit before reaching its operating temperature.

(ii) The telescope temperature is largely determined by the temperature of the outer shell of the payload module. This in turn will vary if the attitude of the spacecraft relative to the sun changes as new pointings are made; careful design will be required to hold the telescope temperature changes within tolerable limits.

(iii) Additional cooling for detectors and instrument optics is needed at all but the shortest IR wavelengths.

(iv) A long mission lifetime implies that degradation of surface properties is likely; since these properties determine the cooling efficiency of the design, it needs to be robust if the long lifetime is to be exploited.

3. How cold can it *really* get?

3.1. Designs and model predictions

A number of models have now indicated that usefully low temperatures can indeed be achieved by the combination of optimised design and optimum orbit. Similar results are rumoured to have been obtained in studies for military projects.

(i) Thermal studies of a warm-launched early design for IRIS (T.Matsumoto, personal communication) in its IRAS-type (900 km sun-synchronous) orbit, with $T_{\rm shell} \simeq 200\,{\rm K}$, predicted $T_{\rm tel} \simeq 37\,{\rm K}$ from purely radiative cooling, with a semi-optimal design (some cooling of individual radiation shields).

(ii) A Cornell/Goddard SFC/SAO design study for SIRTF in December 1993, based on COBE dewar parameters after cryogen boiloff (a non-optimum design, with no cooling of radiation shields) and a 1 m telescope predicted $T_{\rm tel} \simeq 34\,{\rm K}$, or less in a distant orbit.

(iii) Models of a large radiatively cooled telescope (the RCIRT) studied at NASA Marshall Space Flight Center predicted final $T_{\rm tel} \leq 37\,{\rm K}$ (Hawarden et al., 1992) in a 100,000 km circular orbit (the HCO).

(iv) The *Edison* proposal to ESA's M3 opportunity (see Thronson et al., 1994: this volume) built upon (iii) and well illustrates the advantages of a non-geocentric orbit. The *Edison* design, in orbit around the anti-sunward Earth-sun Lagrangian point L2, uses radiation shields cooled by large individual radiators. Spreadsheet and multi-node models for purely radiative cooling predict $T_{\rm tel} \simeq 22\,{\rm K}$ for $T_{\rm shell} \simeq 77\,{\rm K}$.

(v) Lin (1991; also quoted by Rapp, 1992) developed an outline design for a radiatively-cooled 10 m telescope in HCO. This was similar to the RCIRT designs described by Hawarden et al. (1992). The predicted $T_{\rm tel}$ was $\simeq 20\,{\rm K}$.

3.2. Demonstrated on-orbit temperatures

In the open literature we have two practical examples to support the above estimates of how cold a radiatively-cooled telescope can get. Both are provided by missions which were launched as cryogenic telescopes. Other (still classified) examples may exist.

The IRAS telescope was by no means optimally designed for radiative cooling: its radiation shields had no independent cooling and the telescope tube had, by design, low longitudinal conductivity. In its 900 km sun-synchronous orbit its outer shell temperature was $\simeq 200\,{\rm K}$. Following cryogen exhaustion the telescope was parked so that it faced deep space; system temperatures on the spacecraft were measured several times over the next three years (Mason, 1988). That of the telescope optical system was inferred to be close to 100 K, only half the absolute temperature of the outer shell, notwithstanding its non-optimal design.

Further encouragement is offered by the COBE mission. This was better protected from Earth and solar heating than was IRAS: although in a similar orbit, its surface temperature remained around 140 K. After cryogen exhaustion the DIRBE instrument continued to operate, with the detectors (not the coldest parts of the instrument) stabilised at about 43 K *by radiative cooling alone* (Volz et al., 1992, and S.M.Volz, personal communication, 1994).

Thus we have in COBE a *non-optimised* facility operating in a *near-Earth* orbit which achieved radiatively-maintained instrument temperatures only 31% of its absolute outer-shell temperatures. An *optimised* design should clearly achieve a greater fractional reduction. A mission in *deep space* (or at least in an orbit where it is unaffected by Earth radiation) should achieve much lower outer shell temperatures. For example, the shell temperatures predicted for a COBE cryostat in a SIRTF study of December 1993 drops from 140 K to 65 K on moving from the IRAS to a lunar-altitude orbit. If a shell temperature of < 80 K can be achieved, as is predicted by SIRTF studies (Garcia, 1993; Werner & Simmons, 1994: 77.1 K and 60 K, respectively, in heliocentric orbit), a proportional reduction in temperature from outer shell to instrument similar to that occuring on COBE would imply that $T_{\rm tel} \leq 25$ K and possibly between 24 and 19 K could be achievable (from the SIRTF predictions). This naive analogy indicates achievable telescope temperatures in close agreement with the predictions of the models quoted earlier.

4. Combining cooling methods

4.1. Hybrid-cooled IR missions now under consideration

Most of the missions now under discussion actually utilise hybrid approaches to cooling, with varying emphases on radiative, mechanical and cryogenic techniques. The precursor to these designs is the ISO pre-Phase A study by ESA in 1980, which proposed on-orbit radiative cooldown of the telescope after a warm launch and subsequent cryogenic cooling, though in predicting $T_{\rm tel}$ "< 20 K" its authors apparently did not appreciate just how cold the latter could make the telescope!

The Japanese IRIS mission (Matsumoto, 1994: these proceedings) will be launched cold. On orbit the mission will use radiation to cool its outer surfaces and radiation shields, a 2-stage Stirling-cycle refrigerator to cool the intermediate radiation shield, and cryogen boiloff to bring the telescope, again, down to a few kelvins. In an earlier, less conservative design (T.Matsumoto, personal communication), a warm launch and on-orbit cooldown was proposed, similar to that now suggested for SIRTF (see below).

In its most current design (Werner & Simmons, 1994: these proceedings), the SIRTF telescope achieves a rapid initial cooldown after its warm launch by being heat-switched to the payload module outer shell. After the telescope is cold enough ($\simeq 60$ K) this link is opened. The instruments are enclosed in a cryostat containing $\simeq 200$ l of liquid helium. After the shell switch is opened, boiloff from the cryostat is used to cool the optical system down to a few K for operations.

The *Edison* design does not currently include cryogens, but the baseline design (Thronson et al., 1994: these proceedings) has six closed-cycle coolers, three operational systems and a backup for each. These coolers are of the family developed at Oxford and at the Rutherford Appleton Laboratory (RAL) (Bradshaw & Orlowska, 1994: these proceedings), some of which are being marketed by British Aerospace (BAe).

The *Edison* telescope optical system is thermally connected to the telescope tube, while the instrument bay is a separate, thermally isolated, system, heat switched to the telescope for the initial cooldown. The telescope cools radiatively to about 30 K in $\simeq$ 100 days. Main and backup 2-stage Stirling-cycle coolers are then switched on. These bring the telescope to its operating temperature of $\simeq$ 16 K, when the backup cooler is turned off. The instrument bay is then disconnected from the telescope and its own 3-stage ^{4}He Joule-Thompson coolers are turned on. When these have brought the instrument bay to its operating temperature of $\simeq$ 4.5 K, the backup is turned off. At the last stage the detectors are cooled to their operating temperature by ^{3}He 3-stage systems. The use of main and backup closed-cycle coolers in this way more than halves the overall cooling time.

4.2. Hybrid cooling and durability of the *Edison* mission

One of the main potential advantages of radiatively-cooled missions is a long lifetime relative to missions whose duration depends on cryogens. However, simply ensuring that consumables do not set a hard lifetime limit will not suffice to ensure long life. The cooling of the spacecraft depends on selective surface properties in appropriate locations: efficient solar radiation rejection by the sunshade, high infrared reflectivities on the radiation shields, and high emissivities for radiators. Surface properties are notoriously vulnerable to on-orbit deterioration; for example, the second-surface-silvered Teflon widely used for solar radiation rejection slowly becomes opaque in the visible with exposure to solar UV. After a few years its efficiency as a solar rejection surface is greatly reduced. The critical reflective surfaces of the radiation shields, while protected from outside effects, are vulnerable to contamination by outgassing from material inside the spacecraft.

The incorporation of the 2-stage coolers in the *Edison* baseline design, while offering a significant and welcome reduction in the telescope operating temperature, has a possibly more significant advantage, illustrated in Figure 1. At $\simeq$ 20 K the energy flows in the payload are measured in mW: a 20 K black-body field has incident energy density $<$ 9 mW. This means that quite small cooling powers can have a large effect.

The cooling power of the 2-stage "20 K" coolers increases linearly with the operating temperature at the cold tip. At the equilibrium temperature

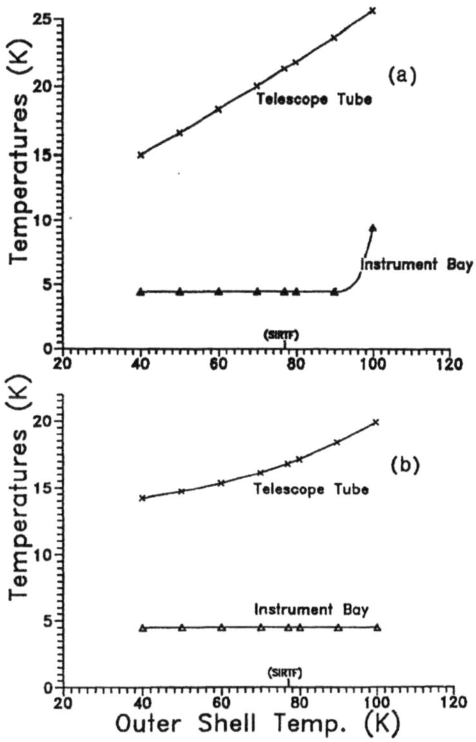

Fig. 1. *Edison* telescope and instrument bay temperatures. The dependence of these on the critical outer shell temperature is shown (a) with closed-cycle cooling (3-stage ^{4}He Joule-Thompson) of the Instrument Bay only and (b) with additional (2-stage Stirling-cycle) cooling of the telescope tube (shell).

of the telescope tube its cooler is lifting only 24 mW via its gas cooling loop, which was modelled at an efficiency of $\simeq$ 50%; but this will increase at 8.3 mW K^{-1} with higher tube temperatures. The tube cooler accordingly has a strong stabilising effect against system temperature rises. As a result T_{tel} rises only to $\simeq$ 22 K, and the instrument bay is expected to remain at its working temperature, for outer shell temperatures up to at least 100 K. The instrument bay temperature is a critical parameter, as quite a small increase will disable the instruments. The 2-stage coolers therefore ensure that the mission is very robust against *e.g.* deterioration of critical surface properties after a long time on orbit. This protection is further assured by the inclusion of a redundant cooler.

5. A Very Cold Telescope for *Edison*

5.1. THE BASELINE TELESCOPE DESIGN FOR *Edison*

The goals of the *Edison* proposal are to provide the astronomical community with a space IR telescope with the largest possible aperture, the longest possible lifetime and the greatest possible sensitivity, in an optimised combination.

As outlined elsewhere, the baseline payload design of *Edison* uses a strong thermal connection between the telescope and its tube, its primary cooling radiator, while (except during cooldown) the instrument bay is thermally isolated, separately suspended on low-conductivity straps within the telescope shell, and protected by its own highly reflective outer cover. In operation the instrument bay is cooled to $\simeq 4.5$ K by a 3-stage ^{4}He Joule-Thompson cooler, another of which is carried as a backup. This temperature is low enough for suppression of internal background at all operating wavelengths and for high-performance detector operation to $\simeq 35$ μm. Beyond this wavelength the detectors are cooled to ≤ 2.5 K by a separate redundant pair of ^{3}He J-T coolers.

The telescope is predicted to stabilize at $\simeq 17$ K in this design. At this temperature it is an extremely powerful facility, but will still generate enough background thermal emission to impair the sensitivity of the mission for spectroscopy at long wavelengths. The telescope is therefore designed like a ground-based IR instrument, with minimum emissivity: it has no baffling and a slightly undersized secondary mirror which forms the entrance stop.

This design, however, may have drawbacks.

5.2. POSSIBLE PROBLEMS WITH THE BASELINE DESIGN

The possible difficulties are, in summary:

(i) Impaired spectroscopic performance at long wavelengths, because of thermal background emission from the "warm" optical surfaces. This can only be eliminated if the optical surfaces are cooled to a few kelvins.

(ii) Secondary mirror diffraction effects, which at long wavelengths give the acceptance cone of the secondary mirror a tapered edge zone. This allows an instrument viewing the secondary to "see" beyond the edge of the primary mirror, even if the secondary is undersized relative to the primary beam. Increasing the degree of undersizing would waste precious aperture. It would be desirable to surround the primary mirror with a ring of highly absorbent surface, cooled to a few kelvins to remove its own thermal emission.

(iii) The unbaffled telescope is vulnerable to stray light from sources close to the direction of observation. This problem will arise even at short wavelengths where the low-background telescope design is not needed. This could be prevented by the use of conventional "optical" type baf-

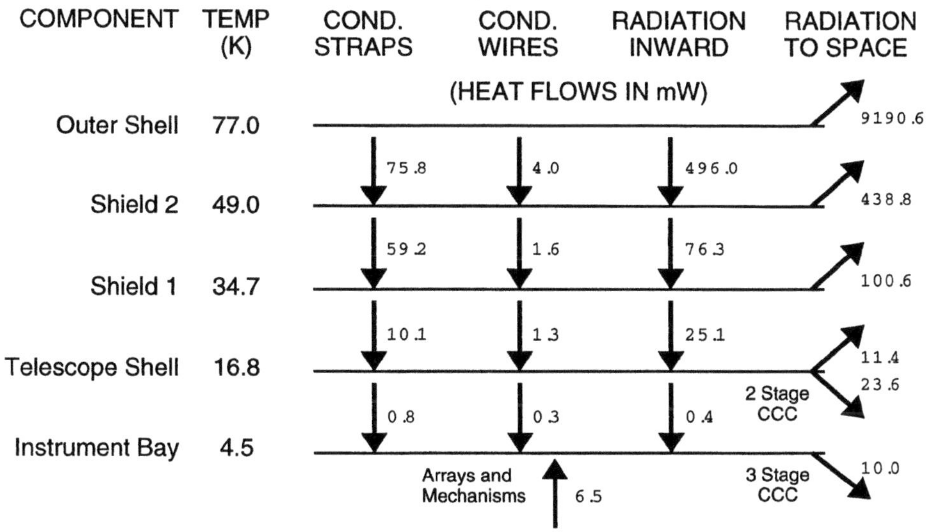

Fig. 2. Heat flows in the baseline design for *Edison*. Note the relatively small heat loads on the telescope shell

fling (see, e.g. the SIRTF design, Garcia, 1993); but again, this would greatly increase the effective emissivity of the telescope and therefore requires that the black baffle surfaces be cooled to a few K.

(iv) Alignment of the instrument bay to the telescope may pose problems if it is a separately suspended system in thermal isolation. Such problems may not be insuperable (recent designs for SIRTF have also adopted this approach) but would clearly be simplified if the telescope and instrument bay were rigidly connected as a single structural entity. This in turn would make thermal isolation of the instrument bay very difficult, and an obvious approach would be to cool the entire structure to the operating temperature of the instrument bay.

5.3. A POSSIBLE SOLUTION.

The foregoing suggests that a means should be sought to cool the entire optical system: mirrors, baffles, absorbing ring and instrument bay, to a few kelvin. Figure 2 shows the heat flows in the baseline design at equilibrium: these are not very much larger than the cooling power available from one of the ^{4}He 3-stage RAL coolers, and suggest that with careful design a fully-cooled structure could be achieved.

5.3.1. Design considerations

The RAL 3-stage ^{4}He coolers can lift about 11 mW at 4.35 K in their present design. The structures we seek to cool are immersed in a radiation field at

≤ 20 K, emitted from the inner wall of the telescope tube, which as noted above has an energy density of only $\simeq 9$ mW m^{-2}; thus if the effective absorbing surface can be kept well below a square metre, an energy balance appears possible. Figure 3 illustrates a possible design for the Telescope Assembly. This includes a fully-baffled telescope, with the secondary mirror supported on a quadripod as the "cleanest" design from the point of view of absorption of IR radiation (as opposed to the design suggested in the M3 Proposal, which uses a half–length metering shell, the cavity of which would be excessively absorptive). The primary is surrounded by a ring of black (highly absorptive) material, 4 cm wide. The primary mirror support structure, its interface with the instrument bay, and the bay itself (now presumed to be firmly bolted to the support structure) are surrounded by a light, highly reflective shroud with a low surface absorptivity.

The whole Telescope Assembly would now be supported by the main straps, inside the Telescope Tube. The inner surface of the latter remains black above the surface of the primary mirror, for efficient emission of radiation during cooldown, and its lower inside surface remains shiny, to minimise inward emission to the TA.

5.3.2. Predicted Thermal Performance

The thermal performance of the proposed design is dominated by two factors: the conductivity of the support straps and the absorptivity of the structure of the Telescope Assembly. Other considerations are parasitic heating by mechanisms and electronics (*e.g.* detectors), and conduction down the electric harness.

(i) Absorptivity of the Telescope Assembly.

- *Black areas comprise*: Secondary baffle, primary hole/baffle, primary outer annulus, backs of legs of the quadripod (visible to the instruments by reflection). *Total black area* ($\alpha \simeq 1.0$) $\simeq 0.4$ m^2.

- *Shiny areas comprise*: Primary mirror surface, instrument bay shell, outer surfaces of quadripod legs, outside of secondary baffle, outside of central baffle. *Total shiny area* ($\alpha \simeq 0.03$) $\simeq 10.9$ m^2.

Total equivalent black area = 0.7 m^2

Heat absorbed from telescope tube/shell (17 K) = $Q_{\rm rad} = 2.3$ mW

(ii) Conduction to the TA and parasitic heating.
Down wires $\simeq 0.3$ mW;
Down CFRP support straps $\simeq 1.7$ mW.
Total conducted heat $Q_{\rm cond} = 2.0$ mW

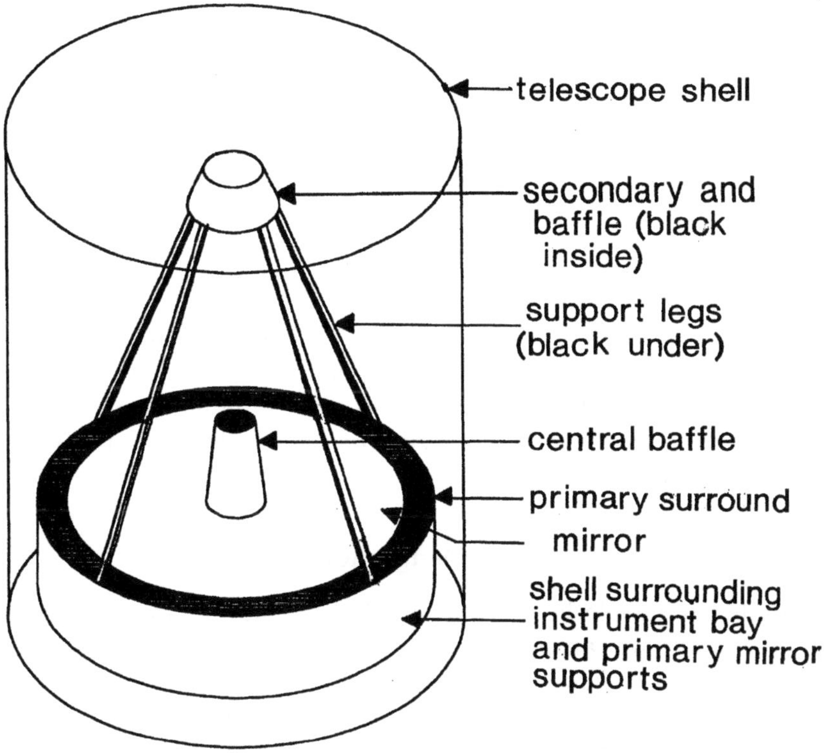

Fig. 3. Telescope Assembly (TA) in the "VCT" option for *Edison* (schematic). The shaded areas have high $\alpha_{ir} = \epsilon_{ir}$, blank areas are "shiny".

(iii) Parasitic Heating by electronics, etc: $Q_{\rm insts} = 6.5$ mW (as in Baseline Design).

To remain at the stabilisation temperature of the ^{4}He J-T coolers, the thermal balance at equilibrium must be:

$$Q_{\rm rad} + Q_{\rm cond} + Q_{\rm insts} \leq Q_{\rm coolers}. \tag{1}$$

The heat fluxes estimated above total to $\simeq 10.8$ mW, a little below the current demonstrated cooling capacity of the RAL ^{4}He coolers, 11.6 mW at 4.35 K (Orlowska & Bradshaw, 1994: these proceedings). Thus even the first-cut design outlined above should be (just) able to maintain this temperature *for the entire optical system and instruments.* Given a somewhat larger margin of safety, whether from futher enhancements of the cooling power of the RAL systems, or refinements of the design, this rough analysis indicates that a long-duration IR space observatory with $T_{\rm tel} < 5$ K could very likely be achieved.

6. Conclusions

We have seen that the use of radiative cooling as a major part of a hybrid system has become almost universal amongst proposed IR space missions since the last Workshop on the Next Generation IR Space Telescope in 1991. Essentially every mission now under serious consideration plans to use radiation as a major component of the process of attaining low temperatures. These are now expected to be low enough that quite small amounts of cooling from ancilliary systems – small cryogen tanks or closed cycle coolers – will suffice to achieve the final step to temperatures of a few K which radiation cannot offer. Hybrid cooling techniques of this type now offer the possibility of a long-lived, large-aperture IR space telescope operating at full efficiency all the way out to the 200+ microns where the submillimetre regime conventionally begins, and where the design of space telescopes begins to be based on different principles.

Acknowledgements

A large number of people have contributed to the development of the ideas outlined herein, and we can single out only a few for explicit thanks. Stephen Volz provided us, *inter alia* with invaluable information on the thermal performance of the COBE cryostat and instruments after helium exhaustion; John Davies, Larry Wade and Don Rapp have encouraged and stimulated the development of these ideas.

References

Bradshaw, T.W., & Orlowska, A.H.: 1994 This conference

Cornelisse, J.W. & Batchelor, M.G.: 1993, "FIRST Rider Study Phase A, Final Report", *DORNIER GmbH*, December 1993

Garcia, M., (ed.): 1993, "Space Infrared Telescope Facility - Mission Concept", *JPL D-11183*, Jet Propulsion Laboratory, October 1993.

Hawarden, T.G., Cummings, R.O., Telesco, C.M. & Thronson, H.A. Jr.: 1992, *Space Science Reviews* **61**, 113

Lin, E.I.: 1991,"A 10-meter 20-Kelvin Infrared Space Telescope: The Passive Cooling Approach",*Proposal to JPL's Director's Discretionary Fund*, Jet Propulsion Laboratory, July 1991

Mason, P.V.: 1988,*Cryogenics* **28**, 137

Matsumoto, T.: 1994, this conference

Rapp, D.: 1992,"Potential for Active Structures Technology to Enable Lightweight Passively Cooled IR Telescopes", *JPL Report D-9449*, Jet Propulsion Laboratory, March 1992

Thronson, H.A.,Jr, et al.,: 1994, this conference

Volz, S.M., DiPirro, M.J., Castles, S.H., Ryschkewitsch, M.G., & Hopkins, R.: 1992, "Final Cryogenic Performance Report for the NASA Cosmic Background Explorer (COBE)",*Advances in Cryogenic Engineering* **37**, ed. Fast, R.W., Huntsville, AL, 1183

Werner, M.W. & Simmons, L.F.: 1994, this conference

THE INFRARED SPACE OBSERVATORY (ISO)

M.F. KESSLER
ISO Science Operations, Astrophysics Division, ESTEC, Noordwijk

Abstract. The Infrared Space Observatory (ISO), a fully approved and funded project of the European Space Agency (ESA), is an astronomical satellite, which will operate at wavelengths from 2.5–240 μm. ISO will provide astronomers with a unique facility of unprecedented sensitivity for a detailed exploration of the universe ranging from objects in the solar system right out to distant extragalactic sources. The satellite essentially consists of a large cryostat containing at launch over 2000 litres of superfluid helium to maintain the Ritchey-Chrétien telescope, the scientific instruments and the optical baffles at temperatures between 2 K and 8 K. The telescope has a 60-cm diameter primary mirror and is diffraction-limited at a wavelength of 5 μm. A pointing accuracy of a few arc seconds is provided by a three-axis-stabilisation system consisting of reaction wheels, gyros and optical sensors. ISO's instrument complement consists of four instruments, namely: an imaging photo-polarimeter (2.5–240 μm), a camera (2.5–17 μm), a short wavelength spectrometer (3–45 μm) and a long wavelength spectrometer (43–196 μm). These instruments are being built by international consortia of scientific institutes and have been delivered to ESA for in-orbit operations. ISO will be launched in September 1995 by an Ariane 4 into an elliptical orbit (apogee 70000 km and perigee 1000 km) and will be operational for at least 18 months. In keeping with ISO's role as an observatory, the majority of its observing time is being made available to the general astronomical community via a "Call for Observing Proposals".

Key words: ISO – Space Astronomy – Infrared

1. Introduction

The scientific potential of infrared astronomy has already been amply demonstrated by observations with ground-based telescopes using infrared instruments operating within the few atmospheric windows, as well as by observations with aircraft- and balloon-borne telescopes. However, all of these systems are limited in wavelength coverage by atmospheric absorption and in sensitivity by the photon shot-noise associated with the thermal emission from the warm telescope optics and from the atmosphere. Operating outside the Earth's atmosphere with a cooled telescope and instruments removes these limitations and gives a major increase in sensitivity.

The first major step in this direction was taken with the highly successful Infrared Astronomical Satellite (IRAS), which surveyed nearly all the sky in four broad infrared bands between 8 and 120 μm. ISO, being an observatory satellite, will build upon the IRAS results by making detailed observations of selected sources. Compared with IRAS, ISO will have a longer operational lifetime, wider wavelength coverage, better angular resolution, more sophisticated instruments and, through a combination of detector improvements and longer integration times, a sensitivity gain of up to several orders of magnitude.

Space Science Reviews **74**: 57–65, 1995.

ISO is designed to be true observatory. It is equipped with scientific instrumentation capable of tackling a wide range of astrophysical problems via imaging and photometric, spectroscopic and polarimetric observations.

2. General Scientific Objectives

Modern astrophysics benefits greatly from the ability to make observations throughout the widest possible range of the electromagnetic spectrum. However, work at infrared wavelengths is severely hampered or totally excluded by the terrestrial atmosphere and therefore, despite its scientific promise, the majority of the $2.5 - 200\,\mu m$ band has remained under-explored.

The ISO part of the spectrum is of great scientific interest, not only because it is here that cool objects ($15 - 300$ K) radiate the bulk of their energy, but also because of its rich variety of atomic, ionic, molecular and solid-state spectral features. Measurements at these wavelengths permit determination of many physical parameters, for example energy balance, temperatures, abundances, densities and velocities. Owing to the much-reduced extinction, infrared observations are particularly well-suited to probing the properties of objects shrouded by dust clouds and thus obscured at visible wavelengths. ISO will be offering high sensitivity and sophisticated observing facilities for a relatively unknown part of the spectrum. Thus it is expected that the scientific programme of ISO will touch upon virtually every field of astronomy, ranging from solar system studies to cosmology.

3. Satellite Design

The ISO spacecraft is 5.3 m high, 2.3 m wide and will weigh approximately 2400 kg at launch. A schematic cut-away drawing is shown in Figure 1. It consists of a payload module, which carries the conical sun shade, the two star trackers and a service module, which provides the basic spacecraft functions. These include the structure and the load path to the launcher, the solar array mounted on the sun shield, and subsystems for thermal control, data handling, power conditioning, telemetry and telecommand (using two antennas) and attitude and orbit control. The last item provides the three-axis stabilisation to an accuracy of a few arcseconds, and also the raster pointing facilities needed for the mission. It consists of sun and earth sensors, star trackers, a quadrant star sensor on the telescope axis, gyros and reaction wheels, and uses a hydrazine reaction-control system. The downlink bit rate is 32 kbit/s, of which about 24 kbit/s are dedicated to the scientific instruments.

The payload module is essentially a large cryostat. Inside the vacuum vessel is a toroidal tank filled at launch with over 2000 litres of superfluid helium, which will provide an in-orbit lifetime of at least 18 months. Some

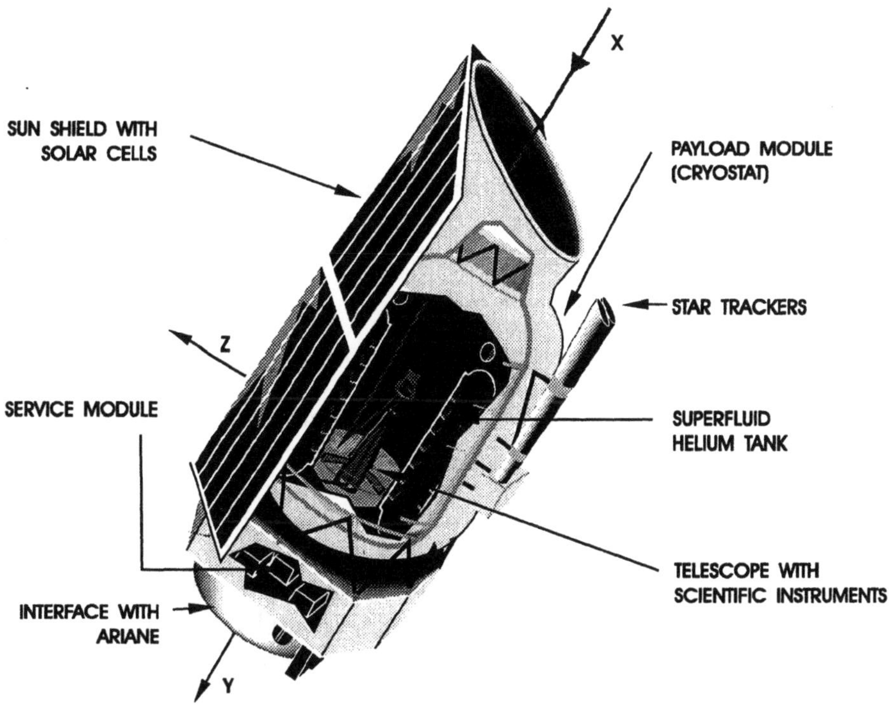

Fig. 1. A Cut-away Schematic of the ISO Satellite

of the infrared detectors are directly coupled to the helium tank and are at a temperature of around 2 K. All other units are cooled by means of the cold boil-off gas from the liquid helium. This is first routed through the optical support structure, where it cools the telescope and the scientific instruments to temperatures of around 3 K. It is then passed along the baffles and radiation shields, before being vented to space. A small auxiliary tank containing about 60 litres of normal liquid helium meets all of ISO's cooling needs for the last 100 hours before launch. Mounted on the outside of the vacuum vessel is a sunshade, which prevents direct sunlight from entering the cryostat.

Suspended in the middle of the tank is the telescope, which is a Ritchey-Chrétien configuration with an effective aperture of 60 cm. The optical quality of its mirrors is adequate for diffraction-limited performance at a wavelength of 5 μm. Stringent control over straylight, particularly that from bright infrared sources outside the telescope's field of view, is necessary to ensure that the system's sensitivity is not degraded. This is accomplished by imposing viewing constraints and by means of the sunshade, the Cassegrain and main baffles, and an additional light-tight shield around the instruments.

The scientific instruments are mounted on the opposite side of the optical support structure to the primary mirror, each one occupying an 80° segment of the cylindrical volume available. The 20 arcmin total unvignetted field of view of the telescope is distributed radially to the four instruments by a pyramid mirror. Each experiment receives a 3 arcmin unvignetted field, centred on an axis at an angle of 8.5 arcmin to the main optical axis, i.e. the instruments view separate areas of the sky. Switching between the instruments will involve a repointing of the satellite.

4. Scientific Instruments

The ISO scientific payload consists of four instruments: a camera (ISOCAM), an imaging photopolarimeter (ISOPHOT), a long wavelength spectrometer (LWS) and a short wavelength spectrometer (SWS). Each instrument has been built by an international consortium of scientific institutes, headed by a Principal Investigator (PI), using national funding and has been delivered to ESA for in-orbit operations. Although developed separately, the four instruments have been designed to form a complete, complementary and versatile common-user package, in keeping with the observatory nature of ISO.

Only one instrument will be operational in prime mode at a time. However, when the camera is not the main instrument, it will be used in a so-called "parallel mode" to acquire extra astronomical data. Whenever possible, the long-wavelength channel of the photometer will be used during satellite slews. This "serendipity" mode will lead to a partial sky survey at wavelengths out to 200 μm, thus complementing the IRAS survey, which only extended to 120 μm in wavelength.

With ISO, photometry and polarimetry will be possible in broad and narrow spectral bands across its entire wavelength range of 2.5 to over 200 μm. A variety of apertures, ranging from 5 to 180 arcseconds, will be selectable out to 125 μm. For spectroscopy, resolving powers ranging from 50 to 30,000 will be available in the wavelength range 2.5 to 196.7 μm. An overview of ISO's photometric and spectroscopic capabilities is given in Figure 2. ISO will be capable of direct imaging in broad and narrow spectral bands across the complete wavelength range at spatial resolutions ranging from 1.5 arcseconds (at the shortest wavelengths) to 90 arcseconds (at the longer wavelengths). In addition, mapping may be carried out using sequences of pointings. Table I gives the main characteristics of ISO's four instruments, which may be summarised as follows:

- **ISOCAM**: a camera for imaging in the 2.5 to 17 μm band with broad-band, narrow-band, polarising and circular variable filters.
- **ISOPHOT**: an imaging photopolarimeter, which contains 3 subsystems for various modes of photometry, namely:

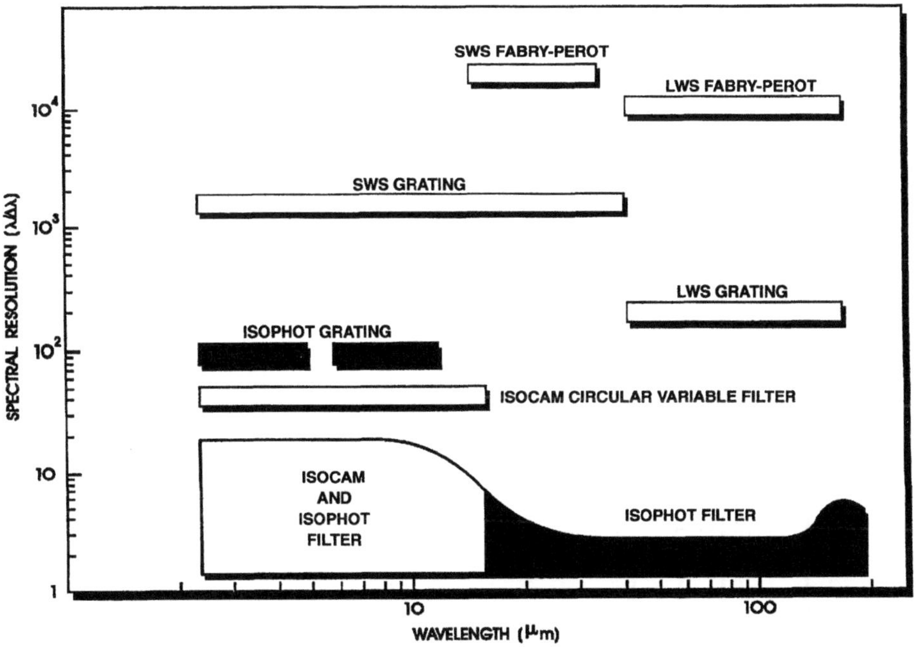

Fig. 2. Photometric and Spectroscopic Capabilities

PHT-P a multi-filter, multi-aperture photopolarimeter for the 3 – 125 μm band,

PHT-C a multi-filter, photopolarimetric camera for the 45 - 240 μm band,

PHT-S two grating spectrophotometers for the 2.5 - 12 μm band (excluding 5 - 6 μm) with a spectral resolution of about 90.

- **SWS**: a short wavelength spectrometer, which consists of two grating spectrometers to cover the 2.4 - 45 μm band with a spectral resolution of about 2000; and with 2 Fabry-Pérot filters to give a resolution of 30000 between wavelengths of 15 and 35 μm.
- **LWS**: a long wavelength spectrometer, which consists of a grating spectrometer to cover the 43 -196.7 μm band with a spectral resolution of about 200; and with 2 Fabry-Pérot filters to give a resolution of 10000 across the same wavelength range.

5. Orbit and Operations

ISO's operational orbit has a 24 hour period, an apogee height of 70000 km and a perigee height of 1000 km. In this orbit two ground stations are needed to provide visibility of the satellite from the ground for the entire

TABLE I

Main Characteristics of the ISO Instruments

Instrument (Principal Investigator)	Wavelength Range and Main Function	Outline Description	Spectral Resolution	Spatial Resolution	Typical Signal (1) Detectable in 200 sec at SNR≥10
ISOCAM (C. Césarsky, CEA-Saclay, F)	2.5–17μm Camera and polarimeter	(i) 32 × 32 array for 2.5–5 μm (ii) 32 × 32 array for 4.5–17 μm	(i) 11 filters 2≤R≤20 circular var. filter R∼40 (ii) 10 filters 2≤R≤14 circular var. filter R∼40	Choice of 1.5, 3, 6 or 12″ per pixel	(i) 3mJy at 2.5–5 μm (ii) 1mJy at 5–15 μm
ISOPHOT (D. Lemke, MPI für Astronomie Heidelberg, D)	2.5–240μm Imaging photo-polarimeter	(i) Multi-aperture, multi-band photopolarimeter (3–110 μm) (ii) Far-infrared camera 30–100 μm: 3 × 3 pixels 100–200 μm: 2 × 2 pixels (iii) Spectrophotometer (2.5-12 μm)	(i) 14 filters 2≤R≤15 (ii) 6 filters 1≤R≤3 5 filters 2≤R≤3 (iii) grating, R∼90	(i) Choice of diffraction-limited to 3′ apertures (ii) 43″ per pixel 89″ per pixel (iii) 24 × 24″ aperture	(i) 30mJy at 10 μm (ii) 15mJy at 100 μm
ISO-SWS (Th.de Graauw, Lab. for Space Research, Groningen, NL)	2.5–45μm Short wavelength spectrometer	(i) Two gratings 2.5–45 μm (ii) Two Fabry-Pérot interferometers 15–30μm	(i) R∼1000 (ii) R∼3×10⁴	(i) 14 × 20″ 14×27″, and 20 × 33″ (ii) 10 × 39″	Grating mode: (2) 6×10^{-16}Wm^{-2} at 12-30 μm
ISO-LWS (P. Clegg, Queen Mary and Westfield College, London, UK)	45–196μm Long wavelength spectrometer	(i) Grating (ii) Two Fabry-Pérot interferometers	(i) R∼200 (ii) R∼10⁴	1′.65 diameter aperture	Grating mode: (2) 3×10^{-16}Wm^{-2} at 55 μm 10^{-16}Wm^{-2} at 100 μm 3×10^{-17}Wm^{-2} at 180 μm

(1) Adopted from corresponding Observer's Manual for each instrument.

(2) Total source flux (continuum plus line) detected in one resolution element.

scientifically-useful part of the orbit – about 16 hours per day. ESA will provide one ground station. ISAS and NASA are contributing to the ISO Project by providing the second ground station and associated resources.

During scientific use, the satellite must always be in real-time contact with one of its ground stations and with its control centre in Villafranca near Madrid, Spain; however, ISO will be operated according to a detailed, preplanned schedule in order to maximise the overall efficiency of the mission. Because of the preplanned nature of the operations, observers have to specify full details of all their observations in advance. Some limited flexibility to respond to unexpected events will be retained in operations.

Various thermal, power, lifetime and straylight considerations place constraints on the allowed viewing directions of ISO. Together the constraints set avoidance regions for ISO around the Sun, the Earth, the Moon and Jupiter. These constraints mean that at any instant only 10 - 20% of the sky is visible to ISO. The orbit precesses relatively slowly; thus, during its entire 18 month lifetime there will be areas of the sky permanently invisible. The location of these areas depend on the date and time of launch. For the planned launch in autumn, the area of zero visibility will lie in the Orion region.

Examination of the scientific data will be carried out both on- and off-line. In close to real-time, a “quick-look” output, adequate for an initial estimate of the success or failure of an observation will be available to the SOC staff. Within a few weeks of an observation being completed, observers will be supplied with data products at various levels of reduction from which they will make their astronomical analyses. Per observation, astronomers will receive a standard set of products (on CD-ROM in FITS format) including re-formatted raw data, reduced images and spectra, calibration data and some specific extracted scientific results to give an indication of what will be found when the observers make their own detailed analysis.

6. Observing Time

Observing time on ISO will be allocated on a “per observation” basis as was the case for EXOSAT and not on a “per night” or “per shift” basis as is the case for many ground-based telescopes and for IUE.

The majority of ISO's observing time is being made available to the astronomical community by the traditional route of “Calls for Observing Proposals”, followed by peer review. One Call has been issued pre-launch and a single Supplemental Call, with restricted access, is foreseen post-launch. The main use of the post-launch Call will be to modify and/or extend and/or redirect existing programmes.

In addition to this “Open Time”, there is “Guaranteed Time” reserved for the groups involved in the preparation and operation of the ISO mission.

Fig. 3. The integrated ISO Flight Model Satellite

These groups are: the four Principal Investigators and their teams, who built the ISO instruments; the five Mission Scientists; the Science Operations Team; the National Aeronautics and Space Administration (NASA), USA; and the Institute of Space and Astronautical Science (ISAS), Japan. A coordinated programme of observations, to be carried out in the guaranteed time, has been prepared by the holders of the guaranteed time and published to the community with the pre-launch Call for Observing Proposals.

The ISO Call for Observing Proposals was released at the end of April 1994 and a total of 1000 proposals was received. Roughly one third of these proposals were for stellar/circumstellar topics, another third for extragalactic studies and one quarter addressed the interstellar medium. The rest of the proposals were split roughly equally between solar system and cosmological subjects. These proposals have been assessed for technical feasibility by members of the Science Operation Centre and and for scientific merit by the ISO Observing Time Allocation Committee (OTAC). Following publication of the OTAC recommendations, successful proposers will be requested to enter full details of their observations into the "Mission Data Base" of the Science Operations Centre in the period up to July 1995.

7. Status and Conclusion

The integration of the Flight Model of the ISO satellite, including the telescope and scientific instruments was completed by the middle of 1994 (see Figure 3. During the autumn of 1994, a wide range of environmental tests – electromagnetic cleanliness, mechanical vibration, acoustic, thermal balance and vacuum – were successfully performed. Compatibility tests between the satellite and its ground segment will be carried out during the first half of 1995. Thereafter, ISO will be shipped to Kourou for its planned launch on 19 September 1995.

Thus, for a period of 18 months in 1995–7, ISO – a sophisticated and versatile infrared observatory in space – will give astronomers the capability to make high sensitivity imaging, photometric, polarimetric and spectroscopic observations at wavelengths from 2 to over 200 μm.

THE INFRARED SPACE OBSERVATORY

Telescope Design

C. SINGER
AEROSPATIALE, Space and Defence Division
100, Bld du midi - BP 99 -
06322 Cannes la Bocca, FRANCE

Abstract. The Infrared Space Observatory (ISO), a programme of the European Space Agency, is an astronomical satellite operating at wavelength from 2.5 to 200 μm. It will be launched in 1995.
The ISO optical subsystem is a cryogenically cooled telescope with its baffling system (main baffle and sunshade). The telescope, a 60 cm Ritchey-Chrétien type, focuses the beam to the four scientific instruments located in its focal plane. The extremely low temperature, 1.8 K, is provided by the payload module (PLM) cryostat, filled with superfluid He.
This paper presents the main choices done for the telescope design together with their rationale and the performances achieved on the flight model (FM) of the telescope. The FM telescope is presently installed inside the payload module, ready for the system final verifications.

1. Introduction

The Infrared Space Observatory (ISO), a programme of the European Space Agency, is a scientific satellite which will permit to perform astronomical observations in the infrared spectrum from 2.5 to 200 μm. ISO will be launched by an Ariane 44p into a highly elliptical orbit with a 24 hours period, 1 000 km perigee and 70 500 km apogee. ISO mission is foreseen to last 18 months.

ISO (Figure1) is composed of two modules, the service module (SVM) and the payload module (PLM). The PLM comprises the cryogenic subsystem, the optical subsystem (telescope, main baffle and sunshade) and the cryoelectronics. The scientific payload is composed of four instruments mounted in the telescope focal plane:

ISOCAM: for imaging and polarimetry in the 3-18 μm wavelength band,
ISOPHOT: imaging and photopolarimetry in the 3-200 μm band,
SWS: spectrometry in the 3-45 μm band,
LWS: spectrometry in the 45-180 μm band.

The telescope and the main baffle surrounding it are located inside the cryostat. They are actively cooled by heat exchange with the cold He-exhaust gas from the He tank. Their low temperatures limit the straylight induced by thermal self emission. At the entrance of the cryostat, a gold coated

Space Science Reviews **74**: 67–72, 1995.

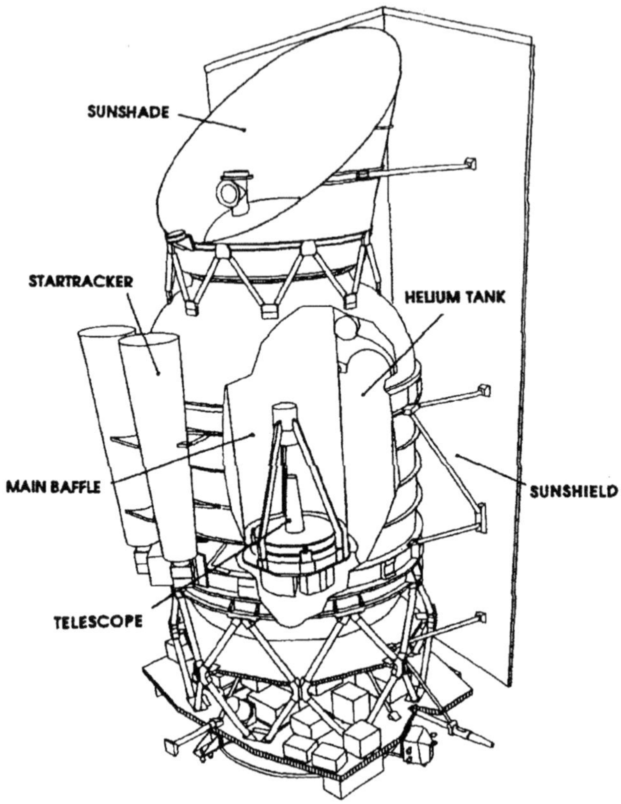

Fig. 1. ISO satellite

sunshade, cooled to 100 K, reject back to space the straylight coming from the Earth.

The telescope, a 60 cm Ritchey-Chrétien type, provides a 20 arcmin unvignetted field of view, splitted by a pyramidal mirror, to the four scientific instruments.

The main baffle, and the cassegrain baffles are black painted, they limit the straylight coming from external bright sources and sunshade thermal self emission.

The overall project is managed by ESA and contracted to an industrial consortium which is led by AEROSPATIALE as Prime contractor. DASA is in charge of the PLM and has subcontracted the optical subsystem to AEROSPATIALE. Each scientific instruments have been provided by a consortium of institutes and industries.

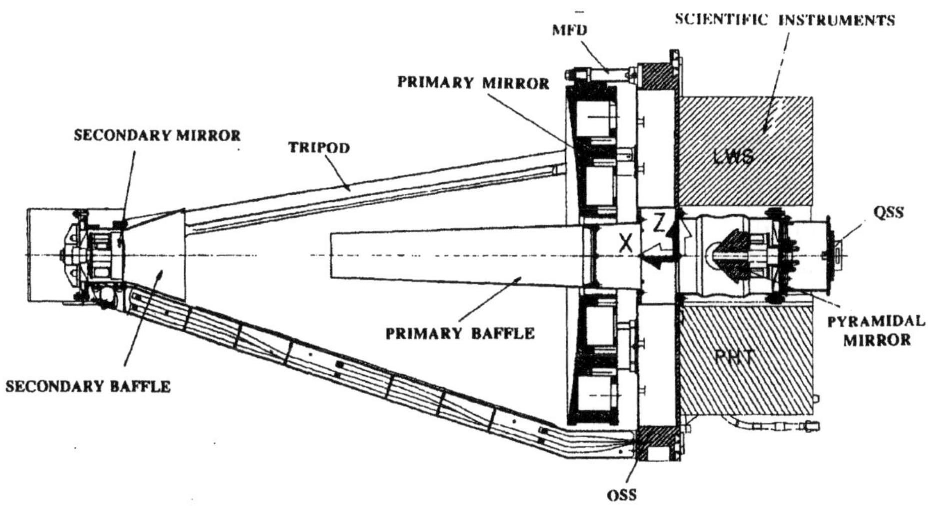

Fig. 2. Telescope design

2. Telescope main requirements

The telescope is a Ritchey-Chrétien type with the following main characteristics:

- Focal length: 9 000 mm
- Entrance pupil diameter: 600 mm
- Exit pupil located on the secondary mirror
- Unvignetted field of view: 20 arcmin
- Wavelength range: 2.5 to 200 μm
- Image quality: diffraction limited at 5 μm
- Straylight: the off axis rejection shall be below 10% of the zodiacal background, idem for thermal self emission.
- Derived from the thermal self emission requirements, the temperature limits for the telescope is 3.2 K (primary mirror and optical support structure).

3. Telescope design

The telescope (Figure 2) is built around the Optical Support Structure (OSS) which supports and cools the three mirrors and the four scientific instruments mounted at its bottom.

The OSS is made of aluminium alloy to have a good thermal conductivity: the He gas evaporated from the He tank circulates inside a loop machined at its rear side, and then goes to the main baffle. The tightness has been measured better than 10^{-8} torr.l/s at 4 K.

The three mirrors are made of fused silica because this material presents a lot of advantages:

- good homogeneity, which is mandatory to avoid surface deformation in cold conditions,
- no hysteresis after thermal cycle,
- low thermal expansion coefficient,
- easy to polish with a good roughness,

These mirrors are gold coated to provide a good reflection coefficient in the infrared, this coating allows also to perform the optical verification in the visible wavelength: WaveFront Error (WFE) measurements and alignment measurements. Through the various cryotests performed on samples (Ungar et al., 1989) and telescope models, we demonstrated the good performances of this coating at cryogenic temperature and after thermal cycles. The disadvantage with fused silica is its low thermal conductivity: to limit the cool down duration of the mirrors, it has been necessary to define thermal straps between the OSS and the primary and secondary mirrors. These straps (OFHC copper) are screwed on the OSS at one side, on the other side, they are screwed on an invar intermediate element (called end bushing) which is glued at the mirror rear face on dedicated silica "finger". Invar has been selected for these end bushing because its thermal expansion coefficient is close to the silica one. But even with this material, it has been necessary to limit as far as possible the gluing surface to avoid silica flakes at cryogenic temperature.

These thermal straps (114 on the primary mirror and 3 on the secondary mirror) have demonstrated their efficiency during the various cryotests: the primary mirror is cooled below 10 K in less than 40 hours, and it has been noticed no optical surface deformation (Collaudin and De Sa, 1988).

The primary mirror is fixed on the OSS via three fixation devices located at the periphery of the mirror 120 degrees apart. To limit the mirror deformations induced by these fixations at 300K and at 4 K, the fixation is made with:

- an invar pad fixed on the mirror and attached in the plane of the neutral fibber of the mirror,
- a flexible link (MFD) between the pad and the OSS which provides the needed degrees of freedom using crossed blades. To take into account the differential TEC between the mirror and the OSS, a pre-loading is applied on the MFD at 300 K.

This type of fixation requires to perform the optical verification with telescope optical axis horizontal, it is not necessary to use an off loading device on the primary mirror to compensate the gravity effect. A zero G device is fixed only at the level of the secondary mirror housing.

A tripod, fixed on the OSS, support the secondary mirror and ensures its positioning stability. It is made with invar to limit the variation of distance between the secondary and primary mirrors, and therefore to limit the telescope focus shift during cool down. A specific thermal treatment is applied during the machining of the tripod to provide a good stability.

Cassegrain baffles are designed to provide a full baffling system. They are made in aluminium alloy, covered with black paint(Ungar et al., 1989). This paint, 500 μm thick, is very efficient even at long wavelength; it is also applied on tripod and on the main baffle which surrounds the telescope.

The telescope adjustment is done at 300 K with ground support equipment (GSE). There is no cryogenic adjustment, which simplifies the design, but requires detailed analysis to get a good behaviour of the 300 K-4 K effect, and to guarantee the final performances of the telescope at 4 K. The adjustment is done in three steps:

- primary mirror adjustment on the OSS (centring and tilt), WFE measurement at 4 K. The accuracy of this adjustment drives the adjustment ranges for secondary and pyramidal mirror,
- secondary mirror adjustment w.r.t the primary mirror (the adjustment is done with GSE, the structure allowing the necessary movement and locking). WFE measurement at 4 K.
- pyramidal mirror alignment to set the four instrument foci at the correct position. Alignment verification at 4 K (at 300 K the secondary mirror is defocused to take into account the thermal shift).

This alignment is performed with instrument alignment dummy which are also representative for thermal and structural aspect.

After delivery by REOSC of the mirrors mounted on their supports, the telescope integration has been performed in class 100 in AEROSPATIALE premises; the cryotests have been performed in Centre Spatial de Liège (CSL), an ESTEC coordinate facility.

4. Telescope performances

The FM telescope has been delivered in September 1993 to DASA for integration inside the FM PLM. The measured performances are the following:

- Image quality: $\lambda/30$ RMS for $\lambda = 5\mu$m,
- Alignment of the four scientific instruments w.r.t the telescope:
  - focus shift: less than 1 mm

lateral shift: less than 0.25 mm
optical axis tilt: less than 7 arcmin

5. Conclusion

The very good performances measured on the FM telescope, confirm its whole design. These results have been achieved thanks to the AEROSPATIALE experience developed on previous optical space programmes, which has been adapted to the cryogenic constraints. With the completion of this programme, AEROSPATIALE demonstrates its capability to design and develop large cryogenic optical subsystems for far infrared applications.

References

Collaudin, B. and De Sa, L.: 1988, "Thermal and cryogenic aspects of the ISO optical S/S", in *3rd European Thermal Control Symposium*, 3-6.10.1988 ESTEC.

Ungar, S., Mangin, J., Jeandel, G. and Wyncke, B.: 1989, "Infrared mirror coating for room and cryogenic temperatures", SPIE, vol.1157.

Ungar, S., Mangin, J., Lutz, M., Jeandel, G. and Wyncke, B.: 1989, "Infrared black paints for room and cryogenic temperature", SPIE vol.1157.

INFRARED TELESCOPE IN SPACE : IRTS

T. MATSUMOTO
Department of Astrophysics, Nagoya University Chikusa-ku, Nagoya Japan

Abstract. The IRTS is a first Japanese infrared satellite mission which will be launched on February of 1995 by HII rocket. The IRTS is one of the mission experiments aboard the small space platform, SFU. The telescope aperture of the IRTS is 15cm, but is cooled by liquid Helium to realize very low background condition. Four instruments are installed on the focal plane which cover wide wavelengths from near infrared to submillimeter regions. The IRTS is optimized to observe the diffuse extended emission, and will survey about 10% of the sky in 20 days of mission life. The IRTS will provide significant information on cosmology, interstellar matter, late type stars, and interplanetary dust.

1. Introduction

Space observation provides excellent environment for infrared astronomy. Owing to the absence of the atmospheric absorption, observations for wide wavelength range can be possible. With use of cold telescope, background radiation of the instrument and telescope itself becomes negligible, and considerably high sensitivity can be achieved.

After the success of the IRAS, importance of the space infrared mission has been widely known. As a next step, some space missions have been proposed. NASA launched the COBE on 1988 to observe the cosmic background radiation and diffuse infrared emission, and is now proposing a new SIRTF mission which is scheduled to be launched in the early 2000s. ESA developed Infrared Space Observatory, ISO which will be launched on late 1995.

In Japan, small space experiments have been actively attained with balloons and rockets. As an extension of these experiments, the IRTS mission was planned as a first satellite mission for infrared astronomy in Japan.

The IRTS is one of mission experiments aboard the small space platform, named as Space Flyer Unit, SFU. Since available space and weight are limited, a small telescope cooled by liquid Helium is adopted and diffuse extended sources are chosen as observation targets (Murakami et al. 1994).

The hardware of the IRTS was completed and the integration on the satellite is in progress. In this paper, the IRTS mission, design and expected performance of the IRTS system are presented.

2. SFU mission

SFU is a small space platform which will be launched with Japanese HII rocket and retrieved by Space Shuttle. The SFU has an octagonal shape with diameter of about 4m, and weighs about 4 tons. Around the central

Space Science Reviews **74**: 73–79, 1995.

hub, several experiment bay are arranged, and the IRTS occupies one of the segments of the octagon. Figure 1 shows side and top views of the SFU in which the IRTS is shaded.

SFU will be launched on February 1995 with HII rocket from Tanegashima Space Center into the low inclination orbit at 500km altitude. A week after launch, the sunshade will be deployed and the IRTS observation will start.

During the IRTS mission, SFU will be uniformly rotated once an orbital revolution roughly around X axis (see Figure 1) to avoid the entrance of the sunshine and earthshine into the cold part of the cryostat. With this attitude control, about 10% of the sky will be surveyed during the IRTS mission.

3. IRTS system

Figure 2 shows the cross sectional view of the IRTS cryostat. The telescope is cooled by super fluid liquid Helium at below 2 K. The liq. He tank has a 100 liter volume and is suspended by support straps made of GFRP from outer shell. The cryostat has three hold vapor cooling shields between which multilayer insulations are inserted. Evaporated He gas first cools forebaffle, and circulates through the vapor cooling shields. Total heat load on the liq. He tank is ~50 mW, and the holding time of liq. He in space is estimated to be about 40 days.

One of the specific issue for the design of the space infrared telescope is the baffling system. Since the orbit of the SFU is in low inclination and low altitude, it is particularly important to avoid both of the earthshine and sunshine. The sunshield which will be deployed in space reflects sunlight and avoids direct incidence of the sunlight into the aperture shade. The aperture shade has a specular surface and thermally insulated from outer shell. The earthshine enters aperture shade but does not hit the surface of the forebaffle in a good condition. The forebaffle is cooled at about 5 K, and has a metallic and specular surface which reflects the scattered and thermal radiation of the aperture shade. Finally, aftbaffle coated by special black paint absorbs stray light.

The IRTS telescope is a Richey-Chrétien system with an effective diameter of 15 cm and a focal length of 600 mm. The mirrors are made of aluminum whose surfaces are polished and gold plated. The performance of the telescope system at low temperature was tested and confirmed to be satisfactory for the IRTS observation (Onaka et al. 1994).

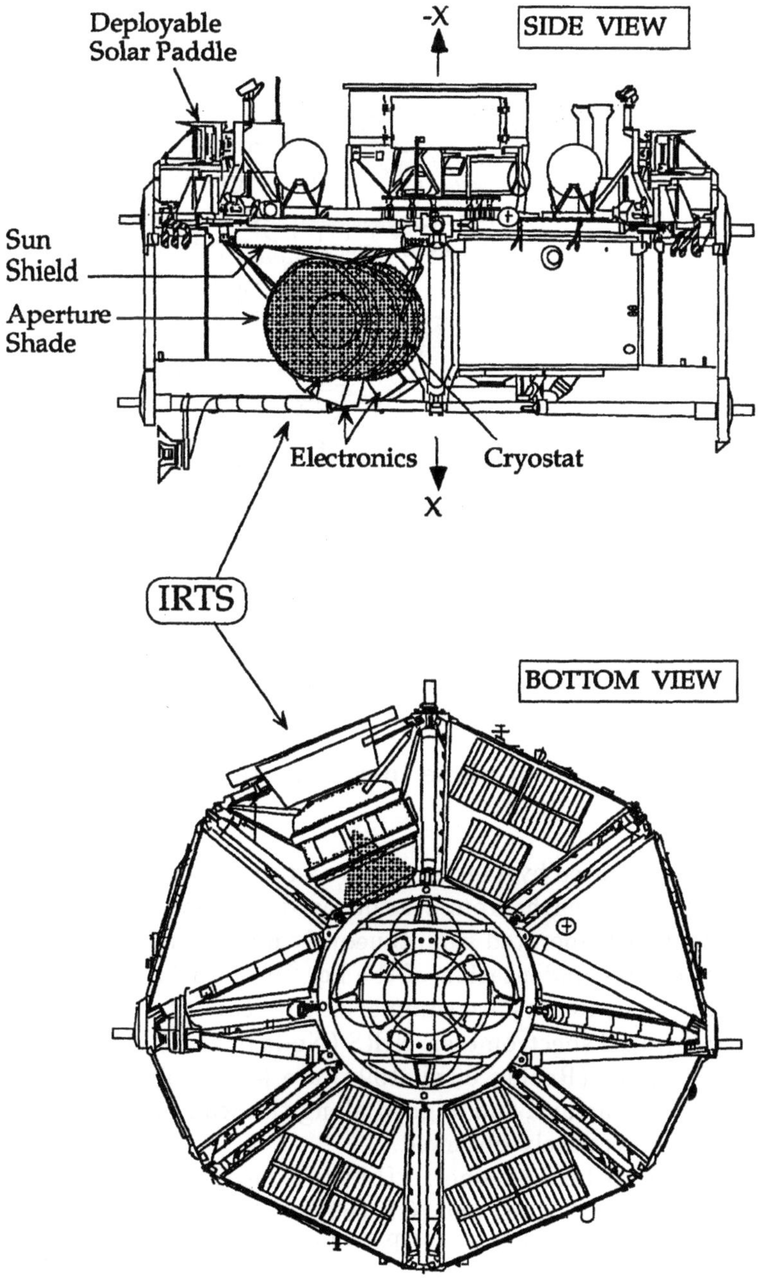

Fig. 1. Top and side views of the Space Flyer Unit, SFU. The IRTS is shaded.

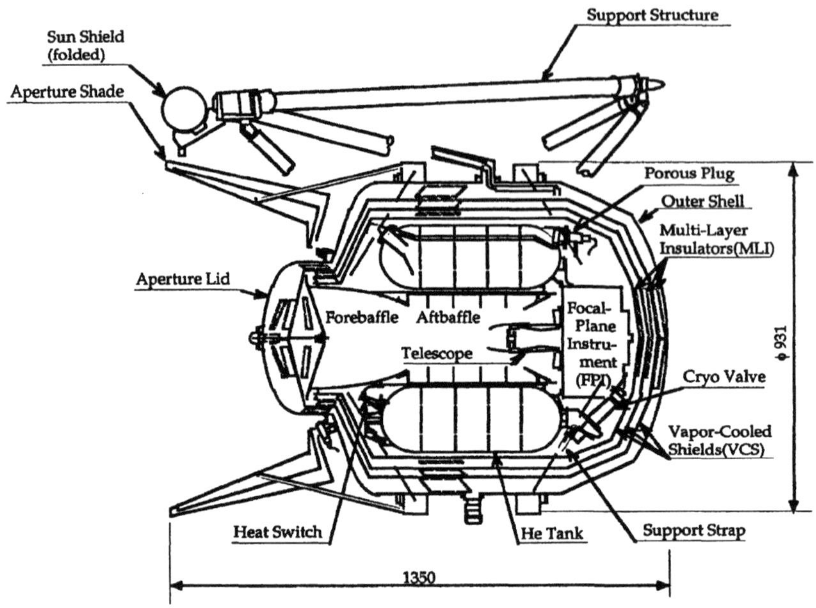

Fig. 2. Cross sectional view of the IRTS. Typical dimensions are indicated in millimeter.

4. Focal plane instruments (FPI)

The focal plane is shared by 4 instruments and a star sensor. The IRTS covers a wide range of wavelengths from near infrared to submillimeter regions. The four instruments are operated simultaneously during the mission, except for the period of the re-condensation of the liq. ^{3}He in orbit. Characteristics of the focal plane instruments are summarized in Table I.

The Near infrared spectrometer: NIRS (Noda et al. 1994) and Mid infrared spectrometer: MIRS (Roellig et al. 1994) are low resolution spectrometers with 1 dimensional arrays, and aligned along the scan path so that the same objects can be successively observed.

The Far Infrared Line Mapper: FILM is a medium resolution spectrometer tuned for [CII] (158μm) and [OI] (63μm) lines.

The Far infrared photometer: FIRP (Lange et al. 1994) is a wide band photometer and designed to observe submillimeter diffuse emission with highly sensitive bolometers cooled by liq. ^{3}He.

The FPI system was calibrated with the test cryostat and it was confirmed that the instruments satisfy the original specification.

TABLE I
Characteristics of focal plane instruments

	NIRS	MIRS
Optical system	grating spectrometer	grating spectrometer
wavelength coverage	1.4-4.0 μm	4.5-11.7 μm
wavelength resolution	$\Delta\lambda = 0.12\,\mu$m	$\Delta\lambda = 0.23 - 0.36\,\mu$m
Beam size	8'×8'	8'×8'
detectors	InSb 24 ch. array CIA, hybrid	Si:Bi 32 ch. array CIA discrete

	FILM	FIRP
Optical system	grating spectrometer	wideband photometer
wavelength coverage	63 μm ([OI]) 158 μm ([CII]) 155, 160 μm (cont.)	140, 230, 400, 700 μm
wavelength resolution	$\lambda/\Delta\lambda = 400$ ([OI], [CII]) $\lambda/\Delta\lambda = 150$ (cont.)	$\lambda/\Delta\lambda = 3$
Beam size	8'×13'	30' circle
detectors	Ge:Ga × 1 Ge:Ga stressed × 3 TIA discrete	0.3 K bolometer × 4 AC biased bridge

5. Astronomy with IRTS

The IRTS is optimized to observe diffuse infrared radiation. In Figure 3, detection limits of the four instruments (hatched pattern) are compared with known diffuse emission sources (solid lines). Thin lines indicate detection limits of the COBE. Detection limits of the NIRS and MIRS are worse than the COBE, however, spectral and spatial resolution of the IRTS are much better than the COBE.

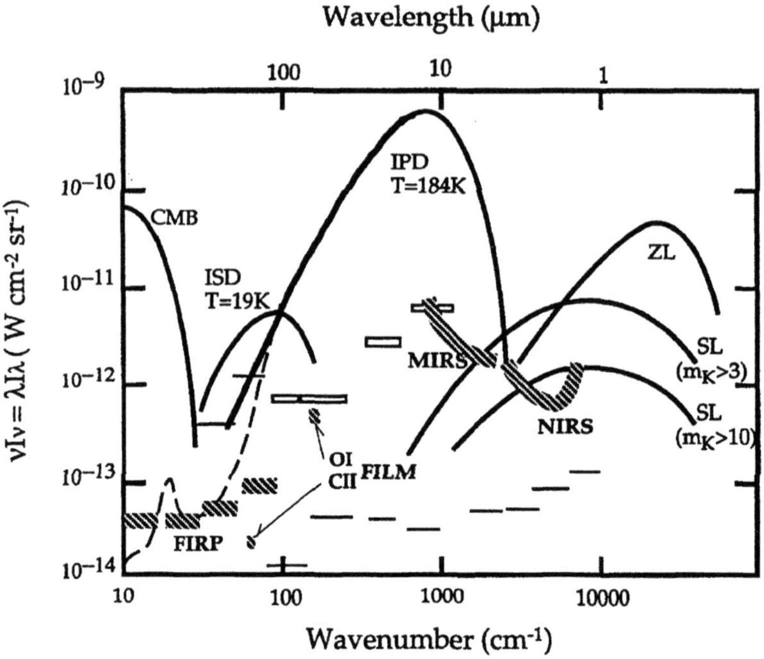

Fig. 3. Detection limits of the focal plane instruments are indicated by hatched pattern. For the FILM, line intensities are plotted. The detection limits are 1σ per field of view. ZL, IPD, ISD, and CMB mean Zodiacal light, thermal emission of the interplanetary dust, thermal emission of the interstellar dust, and 2.74 K cosmic microwave background, respectively.

A variety of observations are expected, however, the followings could be major sciences of the IRTS mission.

As for the solar system, the IRTS has a sufficient power to observe interplanetary dust. The NIRS mainly observes scattered sun light, while the MIRS and FIRP observe thermal emission. Owing to the wide wavelength coverage and the spectroscopic capability, the IRTS will delineate the spatial distribution and the physical properties of the interplanetary dust.

All of the IRTS instruments target the interstellar medium. The FILM will map [CII] and [OI] emission which trace the photo-dissociation region. The MIRS and FIRP observe thermal emission of the interstellar dust. In particular, the NIRS and MIRS can detect extended infrared features located at 3.3, 3.4, 5.6, 6.2, 7.7, 8.6, 9.7, and 11.3 μm due to the PAH.

During the IRTS mission, about 10,000 stars will be detected by the NIRS and MIRS. The IRTS has a wider wavelength coverage and lower detection limits than the LRS on the IRAS. The data set of the stars will be valuable to investigate the structure and evolution of the late type stars.

As is seen in Figure 3, the sky is very dark at 3-4 μm and 200-500 μm. These are windows which enable us to observe extragalactic background radiation. The IRTS is expected to improve our understanding of the infrared extragalactic background radiation.

6. Schedule

After the refurbishment in 1993, the IRTS was tested again, and integrated into the SFU. The SFU will be transported to Tanegashima Space Center by the end of September 1994 and launched on February of 1995.

During the IRTS mission, data will be taken through Deep Space Network (DSN) of NASA. The data will be archived and opened to the public after the initial analysis by the IRTS science team.

References

Lange, A.E., et al.: 1994, *ApJ* **428**, 384.
Murakami,H., et al.: 1994, *ApJ* **428**, 354.
Noda,M., et al.: 1994, *ApJ* **428**, 363.
Onaka,T., et al.: 1994, *Applied Optics* **33**, 1880.
Roellig, T.L,. et al.: 1994, *ApJ* **428**, 370
Shibai,H., et al.: 1994, *ApJ* **428**, 377.

INFRARED ASTRONOMY ON THE MIDCOURSE SPACE EXPERIMENT

STEPHAN D. PRICE
Geophysics Directorate
Phillips Laboratory
Hanscom AFB, MA

Abstract. The Midcourse Space Experiment (MSX) is a multiple objective experiment scheduled to fly by the end of 1994. Infrared photometry and interferometry will be obtained by a solid hydrogen cooled, off-axis telescope of 35 cm unobscured primary aperture. The sensitivities of the line scanned arrays are comparable to IRAS bands 1 and 2 but the spatial resolution is some 30 times better. Nine broadly defined astronomy experiments are planned for the 18 month cryogen phase of the mission. Four of these experiments survey regions not adequately covered by previous infrared missions: the zodiacal cloud near the sun and the anti-solar direction, the Galactic Plane where IRAS sensitivities were limited by confusion and the gaps left by the IRAS survey. The higher sensitivity obtained from raster scans will probe Galactic structure and create intermediate spatial resolution maps of extended sources such as HII regions, the Magellanic Clouds and nearby galaxies. Measurements are also planned on a number of solar system objects such as planets, asteroids, the dust bands, comets and cometary debris trails. Moderate resolution spectra of a number of bright, discrete, extended sources will be obtained as well as low resolution spectral mapping along the Galactic Plane and Zodiacal dust cloud.

1. INTRODUCTION

The next space borne infrared astronomy experiments will be on the Midcourse Space Experiment (MSX) with a launch in the fall of 1994. The mission lifetime is expected to be 5 years, the first year and a half of which will be devoted to measurements with the cryogenic instruments. The remainder of the mission, after the solid hydrogen cryogen has been exhausted, will be given to measurements in the visible and ultraviolet. The circular 898 km altitude orbit is inclined 99.16° and slowly precesses, at $\simeq$7° per year, away from the initial 3:30 am ascending node.

MSX experiment objectives are to collect and analyze target and background phenomenology data to address Ballistic Missile Defense Organization (BMDO, formally the Strategic Defense Initiative) midcourse sensor requirements. The design trades resulting from the different objectives and operating modes limit spacecraft operations to about a 10% duty cycle. Pre-flight mission plans assign an average of two 15-25 minutes astronomical data collection events per day or well over 400 hours of data during the cryogenic phase. The infrared astronomy experiments can be divided into three broad categories: long scans across the sky for large area mapping, raster scans of specific areas to map at higher sensitivity and "stare" mode to obtain spectra with the interferometer.

Space Science Reviews **74**: 81–87, 1995.

A broader description and experiment objective can be found in the paper by Mill et al. (1994).

2. MSX ASTRONOMY INSTRUMENTS

The MSX sensors provide unique measurements capabilities. The instrument complement covers the wavelength range from the ultraviolet to the mid-infrared. MSX will not only obtain simultaneous ultraviolet and infrared images of the sky at about the same spatial resolution but will acquire concurrent spectra over the entire wavelength range.

2.1. Infrared Instruments

The primary instrument on MSX is the Spatial InfraRed Imaging Telescope (SPIRIT) interferometer and a five-band radiometer with scanning arrays of detectors. These instruments share the aperture. Consequently, the interferometer is offset by 2.4° in the scan direction from the principal axis of the optics on which the scanning arrays are centered.

The Michaelson spectrometer collects double-sided interferograms with resolutions of 2, 3.9 or 20 cm^{-1} The interferometer scans at a constant rate so the different resolutions are obtained by selecting the scan times between 4.2, 2.2 or 0.55 seconds which correspondingly changes the scan length. The discrete Si:As impurity band conduction detections are filtered to isolate different spectral regions. Three $6.4' \times 9.6'$ detectors provide spectra over the wavelength 2.6-4.9 μm, 4-26 μm and 2.5-24 μm. The filter for the last channel has a notch at 15 μm to reduce the background from the atmospheric CO_2 Two of the $13' \times 13'$ detectors are filtered for 17.2-28 μm and 5.8-8.9 μm. A slit mask over the third, provides maximum vertical resolution for 11 μm - 13 μm atmospheric measurements which emphasize climate change species such as ozone, nitric acid and fluorocarbons in the stratosphere. The sensitivity of the various channels is roughly $(250/\lambda)$ MJy/sr at 20 cm^{-1} resolution.

The radiometer consists of 5 Si:As integrated arrays of 8 columns by 192 rows each. There is a half pixel offset between the halves of the array. The pixel size is 18″ and the total cross scan field of view is $\simeq 1°$. Dichroics permit simultaneous observations in bands A, D and E and, separately, in B and C. Band B is divided in half with slightly different spectral passbands used for the top and bottom rows. Telemetry limits recording to half of the array columns. Table I lists the spectral bandpasses at 50% response, number of active columns and the estimated sensitivity for each array. (Sensitivity includes the visibility factor: the gain of a matched filter to sample peak. Thus, it includes sampling from two columns, one on either side of the stagger.)

Several ancillary devices are included in SPIRIT III to monitor performance. A shutter can be closed to measure dark offset and noise, emissive

TABLE I
SPIRIT III Radiometer passbands, Array Sizes and Sensitivities

Radiometer Bands	Passband (μm)	No. Columns	Sensitivity (Jy)
Band A	6.0-10.9	8	0.05
Band B1	4.22-4.36	2	5
Band B2			
Band C	11.1-13.2	4	0.2
Band D	13.5-16.0	4	0.2
Band E	18.1-26.0	2	0.3

sources check radiometric stability, a cryogenic quartz crystal microbalance checks for accumulation of condemnable contaminants and an autocollimator measures telescope alignment with respect to the spacecraft optical fiducial to an accuracy of 1″.

Care was taken in the optical design and construction to reduce off-axis radiation. The fore-baffle and internal vanes are painted black to absorb side- lobe radiation. Internal field stops, including an aperture stop which limits the effective entrance aperture to 34 cm, reduce near field diffraction.

2.2. Ultraviolet/Visible Instruments

The infrared instruments are complemented by a suite of nine ultraviolet/visible sensors co-aligned with the radiometer boresight. Narrow and wide field cameras cover the ultraviolet (0.1-0.3 μm) and visible (0.3-0.9 μm). There are also 5 spectrographic imagers (SPIMs) that provide spectra over a two dimensional field.

Telemetry limits data acquisition to a single camera during the cryogenic phase of the mission when SPIRIT III is the prime instrument. We plan to use either the wide or narrow field of view ultraviolet camera. The cameras have complementary spectral ranges, 0.11-0.18 μm for the wide field camera and 0.18-0.3 μm for narrow field instrument. The field of view of the narrow field camera is 1.28°×1.59° which is covered by 20″ pixels. The field and pixel size of the large field camera is about 8.5 times larger.

Collectively, the spectral imagers provide grating spectra from 0.11 μm to 0.9 μm at an average spectral resolution of 175. An internal stepping mirror steps the field perpendicular to the slit. Spectral images of a 1° by 1° field can be mapped in this fashion. Internal binning of the focal plane pixels in both the spectral and spatial directions can be done to increase sensitivity at the expense of resolution in these parameters. All spectral imagers will be used during the cryogen phase of the mission.

2.3. Performance Assessment

The relatively low operational duty cycle and the wide range of measurement geometries means that the instruments and spacecraft will be subject to continually changing conditions A large number of measurement, about 20% of the data collection, will be devoted to assessing and accounting for the changes in instrument performance as the sensor is subjected to varying internal operating parameters, the temperature of the signal processing electronics for example, and those due to the low duty cycle or external influence such as warming of the fore-baffle during an atmospheric measurement. A set of experiments will measure spacecraft effluvia, both molecular and particulate. Such an attention to the details is required to translate the radiometric accuracies available from astronomical sources to the atmospheric observations.

Besides the particles observed around the spacecraft during the primary science measurements, they will be periodically monitored with the large field camera and a xenon flash lamp. Molecular contamination is tracked by a number of crystal quartz microbalances strategically placed around the spacecraft, a pressure sensor and a neutral mass spectrometer. Water vapor concentration is derived by OH emission at 0.305 μm created by dissociation of the water vapor by a krypton flashlamp.

The telescopes have been extensively calibrated in the laboratory with facilities specially built for that purpose. In addition to the usual stellar references, calibration in orbit will also include several standard spheres whose size, thermal properties and spectral emissivities have been accurately measured. A microbalance next to the SPIRIT III primary mirror will measure the amount and rate of cryodeposition. A series of measurements against Jupiter and the Moon track the degradation of the off-axis rejection during the course of the mission.

3. MSX INFRARED ASTRONOMY EXPERIMENTS

The MSX performance parameters are midway between a survey instrument such as IRAS and an observatory, ISO for example, the instrument is ideal for mapping moderate areas of the sky. The MSX pointing constraints are considerably relaxed compared to the IRAS, COBE or ISO. Thus, regions of the sky not accessible to previous or proposed missions are given a priority.

3.1. The Zodiacal Foreground Emission

A priority experiment will be to map the zodiacal light both at near sun elongations and those greater than 130°. The spacecraft does have a thermal constraint not to come within 55° from the sun. Thermal loading from the Earth, however, is acceptable. So the near sun sectors will be scanned using the Earth as an occulting disk on the eclipse side of the orbit. The plan is to twice survey a 20° wide sector in ecliptic longitude to within 25° of the sun. The timing is such that the gap left by the IRAS survey will be mapped concurrently. Five degree wide near sun sectors will be mapped every 5 weeks until the eclipses end.

The zodiacal light at large elongations will be similarly covered. Specifically a $\simeq$40° wide in solar elongation will be surveyed twice as simultaneous redundant maps of the IRAS gaps are made.

All these scans will be almost perpendicular to the ecliptic plane. The detection of linear features aligned along the plane such as the dust bands and cometary debris will be enhance by co-adding across the focal plane.

The IR arrays, an ultraviolet camera, the interferometer and all the spectral imagers will be used to collect data. MSX will obtain low resolution spectra on both the reflected zodiacal light and the thermally re-emitted radiation.

3.2. A Complete and Reliable Mid-Infrared Data Base

Another experimental objective is to combine the MSX measurements with the IRAS and COBE survey products and other catalogs to create and complete and reliable all-sky data base. The survey gaps left by IRAS will be redundantly mapped with half focal plane offsets three time while concurrently surveying the zodiacal cloud. The three redundant maps provide the reliability inherent in IRAS' 3 "HCONS".

Similarly, MSX will map the Galactic plane. The large source density and the IRAS point source extractor rendered the survey reliable only to a density of a source every 25-45 beams. This limited the completeness to fluxes greater than 2-4 Jy within 40° of the Galactic center along the plane. MSX will improve considerably on this. The pixel size is some 30 times smaller than IRAS' and, as part of a preparatory effort, reliable and complete two-dimensional source extraction has been shown to be possible at least to a density of a source every 10 beams. Galactic latitudes within 3° of the plane will be redundantly mapped, the latitude range will be increased to 5° within 60° longitude of the center. A second mapping is proposed if time permits. The observations obviously extent the IRAS measurement. They also complement the COBE near-infrared data with high spatial resolution maps at 4.2 μm maps of a region which is hopelessly confused to the 0.7° COBE detectors. Concurrent infrared spectra along the plane define the diffuse component and provide a data base of mapping the emission in spectral line/bands such as the PAH features.

The anticipated source position accuracy is 2″. These maps provide a valuable resource for locating sources on future missions.

3.3. Raster Scans of Selected Areas

The objectives of these measurements are wide ranging. Maps of a number of regions along the Galactic plane will be made, with emphasis on the longitudes within 45° of the center, create a data base for Galactic structure. Images from the rasters should be about 10 times more sensitive than those from the survey scans.

MSX will map those areas away from the Galactic plane which IRAS labels as confused. Specific regions are the Large and Small Magellanic Cloud and the ρ Ophiuchus, Taurus and Orion molecular clouds... Currently, we plan to survey the regions to improve reliability. Revisits may be done to improve sensitivity.

A number of the infrared bright extended galaxies are to be mapped.

A major issue is what is the structure of the extended emission (Galactic cirrus) like a high resolution. To this end, MSX will conduct a deep map of a high latitude area relatively free of cirrus (Lockman's Hole) as well as the well studied cirrus field near the ecliptic pole.

3.4. Selected Objects

Interferometric mapping at $2\,\mathrm{cm}^{-1}$ of the Galactic center, M17 and the Orion Bar are proposed. While the rather coarse spatial resolution limits what can be obtained, for example it is too coarse to follow shock fronts, the global spectral structure of these regions have not bee mapped over the entire spectral range at this spatial resolution. Additional interferometry at the highest spectral resolution will be obtained on a set of bright "benchmark" HII regions, reflection nebulae, planetary nebulae and AGB stars.

A small number of data collection events will be devoted to solar system objects. Here the unique capabilities of MSX to obtain simultaneous observations over a wide spectral range are important. For example, for comets the spectral signatures of the parent molecules will be observed by the spectral imagers while the daughter species lie in the infrared.

We have a series of measurements which is dedicated to extending the infrared calibration standards to a network of stars more or less uniformly distributed over the sky. The instrument are have been calibrated on the ground with references traceable to NIST standards.

Acknowledgements

The Midcourse Space Experiment is funded by the Ballistic Missile Defense Organization. I wish to acknowledge, in memorium, the foresight and dedicated effort of Dr. Barry which made MSX a reality.

References

Mill, J.D., O'Neil, R.R., Price, S, et al.: "The Midcourse Space Experiment: An Introduction to the Spacecraft, Instrument and Scientific Objectives," *J. Spacecraft*, (in press)

THE WIDE-FIELD INFRARED EXPLORER (WIRE)

P.B. HACKING and M.W. WERNER
Jet Propulsion Laboratory, California Institute of Technology Pasadena, CA 91109

The Wide-Field Infrared Explorer (WIRE) has just been selected by NASA for development under the Small Explorer (SMEX) program, on a schedule leading to launch in 1998. WIRE is a cryogenically-cooled infrared telescope designed to study the evolution of starburst galaxies. It takes advantage of recent advances in infrared detector technology to carry out deep surveys designed to detect distant galaxies at 12 and 25 μm. The WIRE instrument will be integrated with a SMEX spacecraft bus to be provided by the Goddard Space Flight Center and launched into a 500 km orbit by a Pegasus XL launch vehicle.

WIRE will use two 128$\times$128 Si:As IBC arrays cooled by solid hydrogen, which view the same $\simeq$32 arcmin field of view. The telescope aperture is 30 cm. Two principal surveys will be executed during the mission. One will cover several hundred square degrees to limiting flux levels of 1.3 mJy (5σ) at 25 μm and 0.4 mJy at 12 μm. The other, deeper survey, will cover about 20 square degrees to limiting flux levels of 0.4 mJy and 0.1 mJy. This second survey, which will be about 500 times more sensitive than the IRAS survey, is designed to reach the confusion limit at 25 μm. It will probe the Universe to depths at which evolutionary effects are predicted to influence the number and properties of infrared-emitting galaxies. In addition, the catalogs of many tens of thousands of galaxies produced by the WIRE surveys will be a prime source for followup investigations aimed, for example, at discovering additional ultraluminous galaxies similar to those recently identified in the IRAS data bases.

Additional technical and scientific information about WIRE can be found in the following references:

- 'The Wide-Field Infrared Explorer', H. Schember and P. Hacking. *Proc. SPIE, 2019* 395 (1993).
- 'Engineering Design of the Wide-Field Infrared Explorer (WIRE)', D. Elliott, P. Hacking, and H. Schember. Presented at SPIE meeting, San Diego, CA, 25-29 July 1994 (proceedings in press).

Acknowledgements

This research was carried out at the Jet Propulsion Laboratory, California Institute of Technology, under contract to the National Aeronautics and Space Administration

Space Science Reviews **74**: 89, 1995.

SOFIA: THE NEXT GENERATION AIRBORNE OBSERVATORY

E. F. ERICKSON
NASA Ames Research Center

Abstract. The United States and German Space Agencies (NASA and DARA) are collaborating in plans for SOFIA – The Stratospheric Observatory for Infrared Astronomy. It is a 2.5 meter telescope to be installed in a Boeing 747 aircraft and operated at altitudes from 41,000 to 45,000 feet. It will permit routine measurement of infrared radiation absorbed by the atmosphere at lower altitudes, and observation of astronomical objects and transient events from anywhere in the world. The concept is based on 20 years of experience with NASA's Kuiper Airborne Observatory (KAO), which SOFIA would replace. SOFIA will complement the capabilities of other future space missions, and will enable scientists to make observations which would otherwise be made from space.

1. Heritage

For nearly three decades astronomers have been making infrared observations from aircraft based at NASA's Ames Research Center in California. In 1965 Gerard Kuiper used the NASA Convair 990 to show that the clouds of Venus were nearly devoid of water, and demonstrated that aircraft altitudes were suitable for near infrared measurements. In 1968 the Ames Learjet was used by Frank Low to measure far infrared luminosities of Jupiter and Galactic nebulae. In 1974 the Kuiper Airborne Observatory (KAO) started its now 20 year career as a national facility for astronomy. Its users have produced $\simeq$ 1000 refereed publications, and $\simeq$ 50 Ph.D. theses for students at U.S. and foreign universities; it also supports Kindergarten through high school teacher outreach programs. **SOFIA** is planned as the next generation airborne observatory, and is expected to continue this unique program through the first two decades of the next century.

Most infrared radiation from astronomical objects which never reaches the ground is detectable from the lower stratosphere. This fact is the principal justification for an airborne telescope. Figure 1 plots computed atmospheric transmission as a function of wavelength for aircraft (14 km) and mountain-top (4 km) altitudes. The absorption is largely due to water vapor, with significant contributions from carbon dioxide and ozone in some wavelength bands. The model assumes overhead water quantities of 2.3 microns (μm) for the aircraft and 1.2 millimeters for the ground-based telescope, and a zenith angle of 60 degrees. From 14 km, the broadband transmission is adequate for photometric observations at most infrared wavelengths. A number of the water lines in the far infrared are still saturated at aircraft altitudes, as shown in the lower panel. Nevertheless, most important astronomical spectral features can be measured from an airborne observatory.

Space Science Reviews **74**: 91–100, 1995.

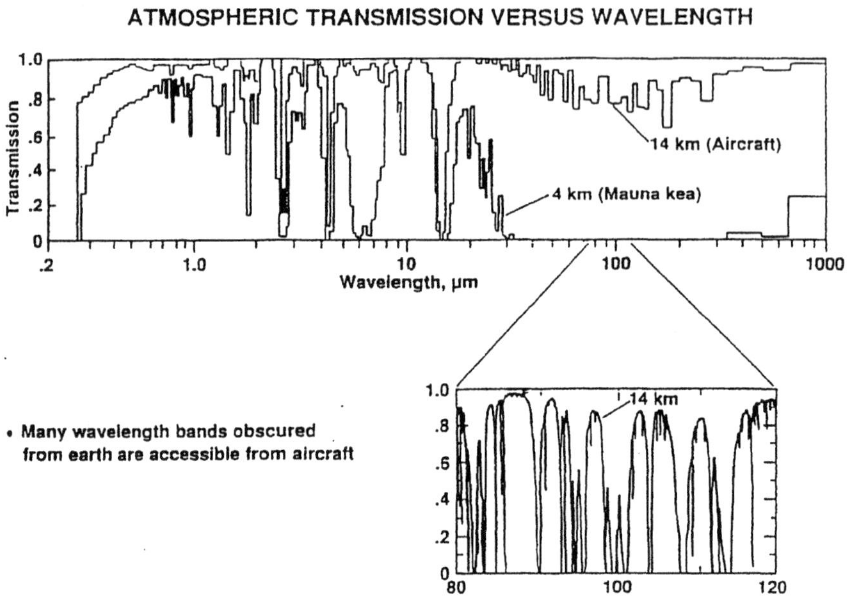

Fig. 1. Calculated atmospheric transmission.

2. Characteristics

SOFIA's characteristics are summarized in Table I. The large aperture and wavelength range, and the routine accessibility to most infrared wavelengths are unique features of the observatory. Relative to the KAO, SOFIA will be roughly ten times more sensitive for compact sources, enabling observations of fainter objects and measurements at higher spectral resolution. Also, it will have three times the angular resolving power for wavelengths greater than about 10 μm, permitting more detailed imaging throughout the far infrared.

Anticipated performance curves as a function of wavelength are given in Figures 2 – 5. Figure 2 shows the anticipated image quality, which is limited by seeing from the air flow over the telescope cavity at visible and near infrared wavelengths, and by diffraction at long wavelengths. The specified performance of the optical system limits the image quality in the $\simeq$ 4–10 micron range.

Figure 3 shows the anticipated photometric sensitivity per pixel, which is simply scaled from the performance achieved on the KAO. Here "PSC" and "FSC" refer to the IRAS Point Source and Faint Source Catalogues, respectively. We see that SOFIA would be able to observe any of the far infrared sources in these catalogues.

TABLE I
Summary of Basic SOFIA Characteristics

In-flight access to focal plane instruments	continuous
Vehicle	Boeing 747
Operating altitudes	41,000 to 45,000 feet
Aircraft-limited duration at or above 41,000 feet	$\gtrsim$ 5 hours
Nominal flight duration (crew-limited)	7.5 hours
Research flights per year	$\simeq$ 160
Number of PI teams flown per year	$\simeq$ 60
Number of focal plane instruments flown per year	$\simeq$ 15
Stabilized telescope system weight	$\simeq$ 8,000 kg
Effective primary mirror diameter	$\simeq$ 2.5 meters
Wavelength range	0.3 to 1600 microns
Telescope configuration	Nasmyth
Design f/ ratio	$\sim$ 20
Design plate scale	4 arcseconds/mm
Unvignetted field of view	8 arcminutes
Telescope optical image quality	$\gtrsim$ 1 arcseconds
Shear layer seeing	$\simeq$ 3 arcseconds (0.3 - $\simeq$5 microns)
Diffraction-limited at wavelengths	$\gtrsim$ 10 microns
RMS pointing stability	0.2 arcseconds
Telescope emissivity	$\lesssim$ 15%
Nominal operating optics temperature	$\simeq$ 250 K

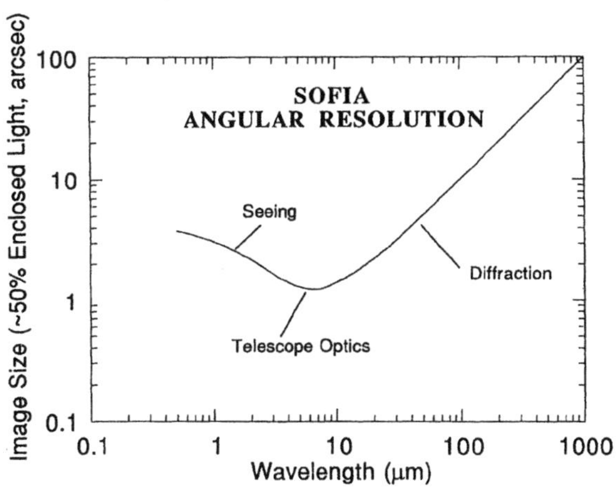

Fig. 2. Angular resolution.

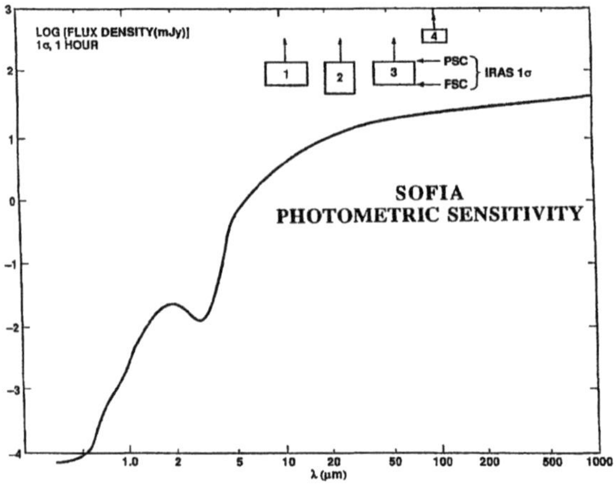

Fig. 3. Photometric sensitivity.

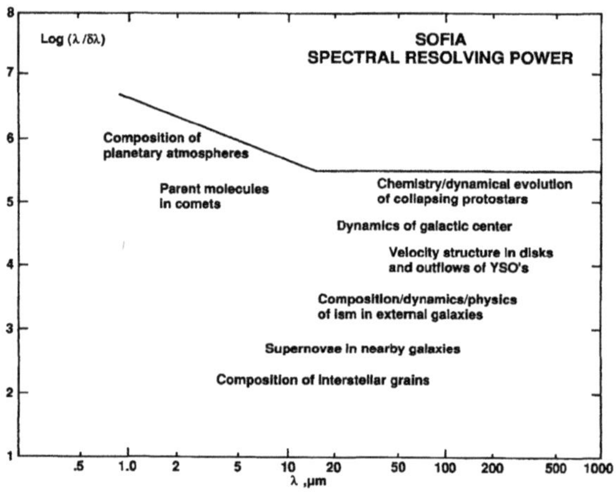

Fig. 4. Spectral resolving Power.

High resolution spectrometers are expected to be available for most of the wavelength range of SOFIA, as indicated in Figure 4. Interstellar lines are typically broadened to a km/s or more, whereas higher resolving power may be required for study of solar system objects.

Spectroscopic sensitivity, shown in Figure 5, corresponds to the spectral resolving power shown in Figure 4. Narrow lines emitting more than 0.1% of the total continuum emission from the IRAS sources should be detectable. The important cooling lines of neutral oxygen at 63 μm and C^+ at 158 μm from photodissociation regions are typically this strong relative to the con-

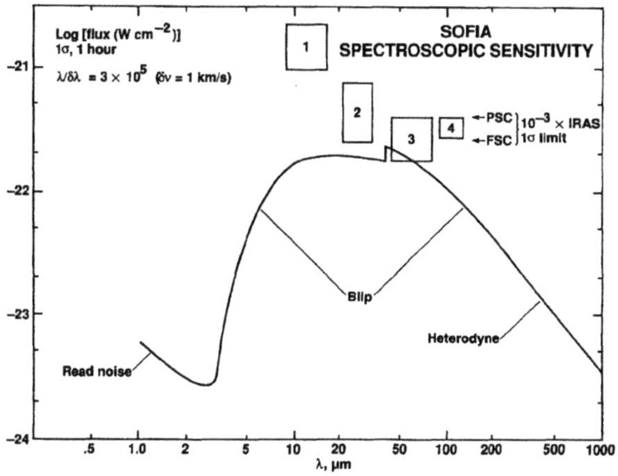

Fig. 5. Spectroscopic sensitivity.

tinuum. Shocked interstellar gas will produce a higher line to continuum ratio.

3. Science

The infrared spectral regime encompasses a multitude of rich and varied physical processes and is uniquely suited for study of the cosmic birth on all scales. SOFIA's wavelength range from $0.3\,\mu$m to 1.6 mm includes this entire regime, and permits a very wide variety of astronomical problems to be addressed. With its high spectral and spatial resolution, SOFIA will exploit and extend the scientific legacy left by IRAS (the Infrared Astronomical Satellite). As seen above, SOFIA will have adequate sensitivity to study any of the band 3 or 4 sources in the IRAS survey. It will complement the spectacular sensitivity for imaging and moderate resolution spectroscopy to be furnished by ISO, the Infrared Space Observatory, and by SIRTF, the Space Infrared Telescope Facility. The National Academy of Sciences has given SOFIA a high priority for development this decade. Topics and observational capabilities are summarized in Table II.

Clearly the diversity and depth of the science program for SOFIA are too extensive to be reviewed with completeness here. Instead we offer only a few examples.

Roughly a quarter of the science on SOFIA will be done on solar system objects. One type of observation which SOFIA will greatly enhance is occultations. The frequency of planets and ring systems occulting observable stars will increase from one every few years (on the KAO) to about one per year, permitting resolution of morphological details on the scale of a few

TABLE II
SOFIA Science

⋆ **Major science topics:**
– Formation of stars and planetary systems
– Physics of the Galactic Center
– Energy sources and ISM in galaxies
– Structure and chemistry of solar system objects
⋆ **Main observational capabilities:**
– Far infrared photometric and polarimetric imaging
– High resolution spectroscopy/spectroscopic imaging
– Occultations, eclipses, novae

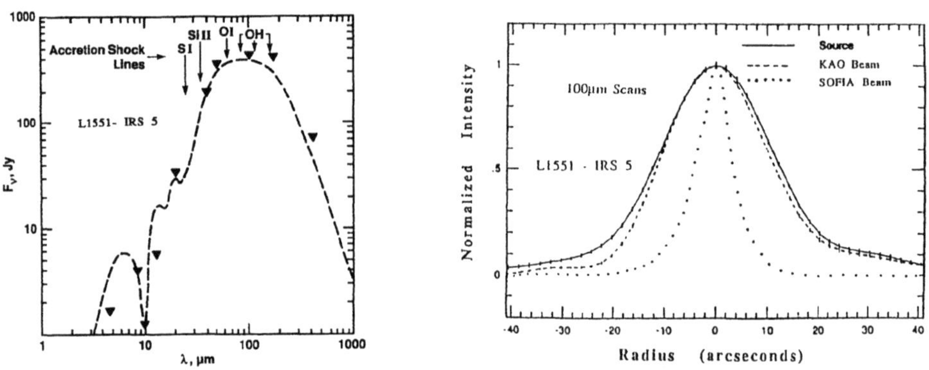

Fig. 6. (a) The IR-Submm spectrum of the YSO L1551 IRS-5. (b) Comparison of SOFIA and KAO resolution compared to the physical size of L1551 IRS-5 at 100 μm.

kilometers. Exciting possibilities are determination of the composition and structure of atmospheres on Pluto and Titan, and of the rings of Saturn and Uranus.

Numerous spectral features will be measured to diagnose the composition, dynamics, chemistry, density, temperature, luminosity, and magnetic fields of the interstellar medium. Many of these features have already been observed from the KAO; these characterize the solid, molecular, neutral atomic, and ionized regions of the ISM. For example, spectroscopy will yield particularly significant results on carbon chemistry in clouds, via the [CII] 158 μm line, the 370 and 609 μm [CI] lines, and a host of far infrared and submillimeter rotational CO lines.

Studies from SOFIA of star formation in the nearby (150 pc) Taurus cloud will reveal the stellar luminosity function for young stellar objects (YSO's) down to values below $0.1\,L_\odot$; most of the luminosity is radiated in the far infrared, where SOFIA's angular resolution is needed to avoid confusion. Figure 6a depicts the spectrum of L1551 IRS-5, a YSO in Taurus. Evidence for extended structure – probably corresponding to cloud collapse around these stars – has already been marginally detected from the KAO. Note that the Jeans length for a $1\,M_\odot$ cloud is $\lesssim 1'$ at Taurus. SOFIA's $8''$ beam (corresponding to $\simeq 1200\,$AU) at $100\,\mu$m could map out the asymmetries of the infall region, which probably accompany the formation of an accretion disk. Figure 6b compares the KAO data with what could be done from SOFIA. Although the disks are probably too small to be resolved by SOFIA, their presence and clues to their character (mass, accretion rate, and dynamics) may be deduced by their continuum spectrum and by the emission of shock-induced far infrared lines of [SI], [SiII], [OI], and OH (Table I). This information will complement the longer wavelength images of the outer (cooler) parts of the disks, which will be obtained with the planned ground-based millimeter arrays. Note that SOFIA can probe the $\simeq 60\,\mu$m wavelength range which corresponds to the region of the disk where planets are expected to form. Finally the common but enigmatic outflows often associated with YSO's can be imaged in the far infrared continuum and lines with SOFIA to examine at what evolutionary stage the outflows "turn on". Far infrared polarimetric imaging of the infall regions from SOFIA will show, in the more luminous cases, the association between magnetic fields and the outflows.

SOFIA will also be an important tool in understanding our own Galactic Center. Suffering a visible extinction $\simeq 10^{10}$ and with an infrared luminosity $\simeq 10^7\,L_\odot$, this unique source has been partially unveiled already by observations from the KAO. Figure 7 illustrates the discovery by KAO investigators of a ring-like structure of cool dust emitting predominantly at $\simeq 100\,\mu$m, which surrounds a low density cavity $\simeq 1\,$pc in extent. In addition, KAO spectral images of the $63\,\mu$m fine-structure line of [OI] revealed the presence of a large amount of neutral gas in this cavity. This gas may be falling into the very center of the Galaxy along what appear to be ionized gas streamers seen at radio wavelengths. The mass of these neutral clouds is much larger than that of the ionized streamers, suggesting that the ionized gas streamers are just the surfaces of these atomic clouds. It is speculated that these surfaces are ionized by powerful but unseen nearby UV sources. At $63\,\mu$m on the KAO, the spatial resolution is about $20''$, leaving unanswered several important questions: What are the origins of this streaming gas? Does it arise locally, from the circumnuclear disk – or, does it come streaming in from clouds tens of parsecs away? Is it truly spiraling into the radio point source Sgr A*? If so, how close to the very center may we trace the gas? If there is a black hole at Sgr A*, when may we expect the next large outburst as these atomic clouds begin to be accreted onto its "surface"? With the

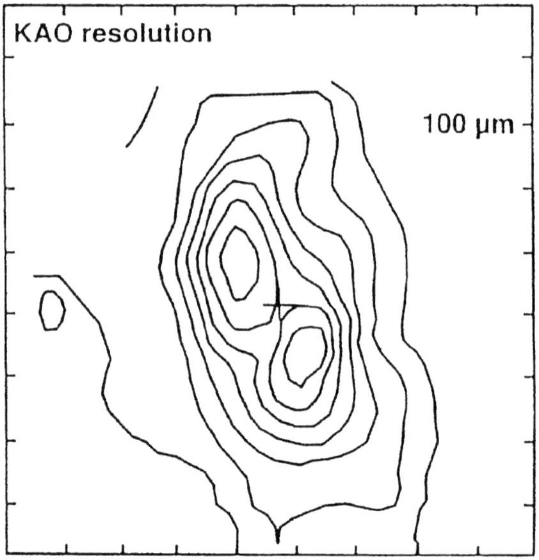

Fig. 7. The Galactic Center seen by the KAO.

factor of three higher spatial resolution afforded by SOFIA, we expect to answer these and many other questions.

SOFIA will explore these and a wide range of other science topics utilizing the unique features of its airborne predecessors: spectral coverage from the near UV to millimeter wavelengths, a routinely accessible "hands-on" working environment, annual opportunities for proposing new science and instrumentation, and deployability for all sky coverage and observations of ephemeral events such as comets, eclipses, occultations and novae.

These attributes, plus SOFIA's sensitivity, angular resolution, and 20 year lifetime, assure a vigorous and productive science program. The broad-based community participation will include extensive opportunities for young scientists and for prompt application of new instrument technologies. Currently 160 8-hour missions per year are planned for over 40 principal investigator teams. Many of the investigator teams will provide the observatory with specialized instruments, including array cameras, polarimeters, and a variety of spectrometers. Thus, in addition to its wide-ranging scientific contributions, SOFIA will stimulate development of focal plane instruments, and provide important educational experience in support of future space astronomy missions.

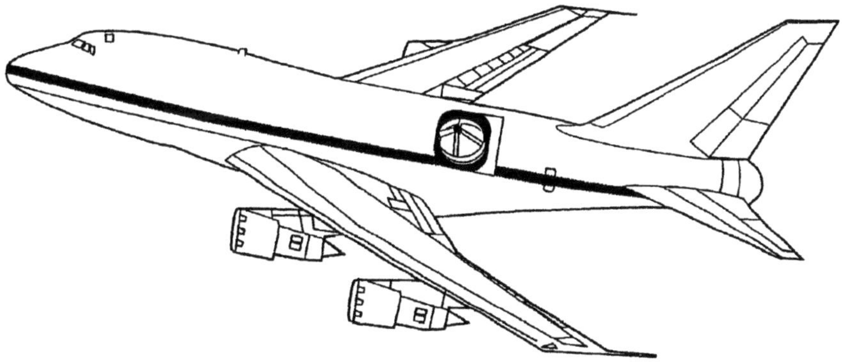

Fig. 8. SOFIA.

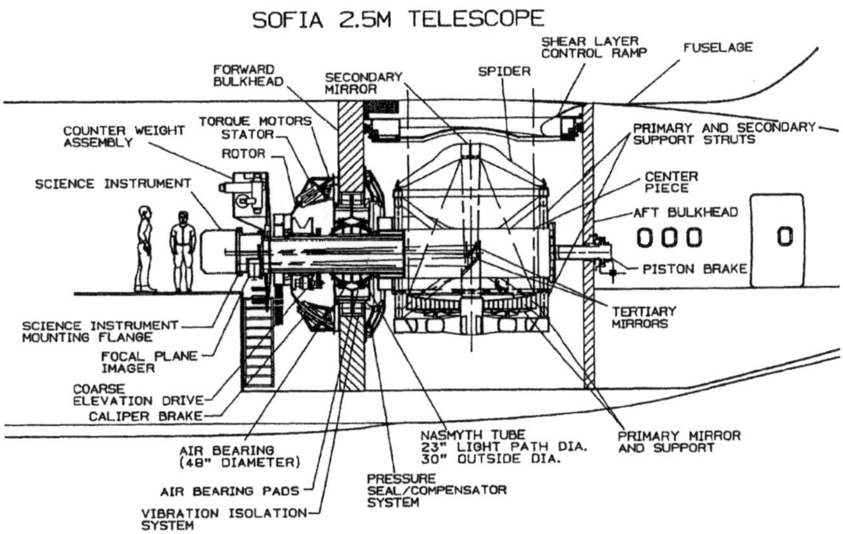

Fig. 9. The SOFIA Telescope Assembly.

4. Readiness

Wind tunnel tests and computational fluid dynamic modelling of the airflow over the open-port telescope cavity have resulted in a quiet, low drag shear-layer control concept for SOFIA. A configuration with the cavity located behind the wing has been shown to be much less expensive than putting the telescope ahead of the wing. This configuration has been adopted, although it increases the "seeing" distortion of the images at near infrared wavelengths (Figure 2) due to the thicker boundary layer in the rear of the plane. Figure 8 shows an outline drawing of the plane with the aft telescope configuration.

The telescope design (Figure 9) features an airbearing support and numerous other similarities to the KAO telescope, which has achieved sub-arcsecond pointing stability even in light turbulence. Structural and optical analyses indicate that either metal or composite structure could be used, and that any of several (glass) primary mirror designs would work. The 2.5 m primary can be as slow as f/1.7 with the telescope in the aft location, so that figuring is not a problem.

5. Status

SOFIA definition studies, sponsored jointly by NASA and the German Space Agency DARA, have been completed, and the project has been deemed ready for development by these agencies. Pre-development studies of the telescope are planned in Germany, and new wind tunnel tests of the aft telescope configuration are planned at NASA Ames. NASA and DARA plan to begin development in 1996 if funding is available.

References

Traub, W.A., Stier, M.: 1976, *Applied Optics* **15**, 364
Erickson, E.F., Davidson, J.A., Caroff, L.J., Thorley, G.: 1991, *NASA TM 103840*

POST: POLAR STRATOSPHERIC TELESCOPE

PIERRE Y. BELY,* HOLLAND C. FORD, RICHARD BURG,
LARRY PETRO and RICK WHITE
Space Telescope Science Institute, Baltimore, MD USA
and
JOHN BALLY
University of Colorado, Boulder, CO USA

Abstract.
The tropopause, typically at 16 to 18 km altitude at the lower latitudes, dips to 8 km in the polar regions. This makes the cold, dry and nonturbulent lower stratosphere accessible to tethered aerostats. Tethered aerostats can fly as high as 12 km and are extremely reliable, lasting for many years. In contrast to free-flying balloons, they can stay on station for weeks at a time, and payloads can be safely recovered for maintenance and adjustment and relaunched in a matter of hours. We propose to use such a platform, located first in the Arctic (near Fairbanks, Alaska) and, potentially, later in the Antarctic, to operate a new technology 6-meter, diluted aperture telescope with diffraction-limited performance in the near infrared. Thanks to the low ambient temperature (220 K), thermal emission from the optics is of the same order as that of the zodiacal light in the 2 to 3 micron band. Since this wavelength interval is the darkest part of the zodiacal light spectrum from optical wavelengths to 100 microns, the combination of high resolution images and a very dark sky make it the spectral region of choice for observing the redshifted light from galaxies and clusters of galaxies at moderate to high redshifts.

Key words: Infrared telescopes – Balloons

1. Introduction

The excellent conditions offered by the stratosphere for infrared observations are well known. The stratosphere is very dry, which minimizes atmospheric opacity and radiance, an advantage magnified by the low pressure which reduces line broadening. The ambient temperature is cold, which minimizes telescope optics emission. Finally, the stratification hampers the generation of turbulence and the associated "seeing" effects. In the near infrared, and to a lesser extent in the longer infrared and submillimetric part of the spectrum, the resulting conditions approach those of space.

Observations in the near-IR from aircraft suffer from image degradation due to air turbulence and vibration. Operational costs are also fairly high. Free-flying balloons reach greater altitudes, but observation time is also limited, and reflights infrequent.

When very high altitude is not absolutely required, as is the case for the 2 to 3 micron band, tethered aerostats come close to the convenience of ground observatories. As opposed to traditional free-flying balloons, tethered

* Affiliated to the Astrophysics Division, Space Science Department, European Space Agency

Space Science Reviews **74**: 101–112, 1995.

aerostats offer reliable launch and recovery of payloads, long on-station time, and the lowest cost per hour.

The tropopause, typically at 16 to 18 km altitude at the lower latitudes, dips to 8 km in the polar regions. It is difficult for a tethered aerostat to fly at altitudes much higher than 12 km. But near the poles, this is high enough to be well into the stratosphere most of the time. Combining the practicality of tethered aerostats with siting in a polar region to benefit from a lower tropopause is the key idea behind our proposed POST concept.

The location of the observatory does not matter much as far as observation conditions in the stratosphere are concerned. This is because the stratosphere, contrary to the troposphere, is essentially decoupled from ground influence. In the Antarctic, conditions in the lower stratosphere are generally more stable than in the Arctic, with lighter geostrophic winds, and temperatures colder by about 20°C. However, conditions on the ground are much harsher and access and logistics are markedly more difficult. Consequently, we propose to first locate POST in the Arctic, possibly near Fairbanks, Alaska, with a possible subsequent relocation to Antarctica to observe important objects in the southern sky.

In what follows, we first describe the excellent observing conditions offered by the lower polar stratosphere, and then we describe the main characteristics and performance of the aerostat and telescope that we propose to build to exploit those conditions.

2. Observing from the lower polar stratosphere

2.1. Atmospheric transparency and radiance

At 12 km altitude in arctic or antarctic regions, the pressure is about 17% of the sea level atmospheric pressure, and the column amount of precipitable water above is only about 10 microns. The corresponding atmospheric transmission and radiance have been estimated using the FASCODE program augmented to include the non-thermal emission in OH lines at the lower end of the spectrum, and are shown in Figure 1.

Except for two bands dominated by H_2O absorption around 2.7 and 4.3 microns, the infrared transmission at 12 km is superb. At wavelengths less than 2.2 microns, both POST and ground-based telescopes see the same bright background dominated by OH emission. At wavelengths longer than 2.2 microns, the radiance POST sees is one to three orders of magnitude less than on the ground. The region between 2.2 and 2.7 micron is of particular interest because the atmospheric radiance is at a lower level than the zodiacal light which is, itself, at a minimum there (30 phot $\mathrm{cm^{-2}\,s^{-1}\,\mu m^{-1}\,arcsec^{-2}}$).

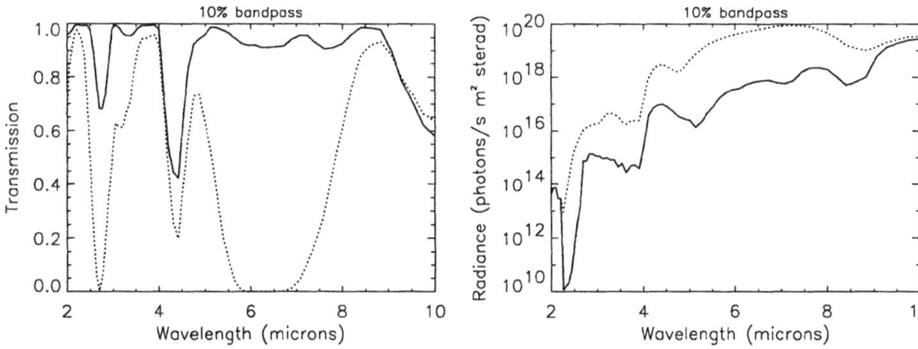

Fig. 1. Atmospheric transmission and radiance at zenith for an excellent ground observatory (Mauna Kea, Hawaii) - dotted line, and at 12 km altitude - solid line.

2.2. Seeing

Atmospheric turbulence which strongly affects telescopic images in the visible (seeing) varies as the $-6/5^{\text{th}}$ power of wavelength and is thus less pronounced in the near infrared. Still, at 2.2 microns, for example, Fried's parameter r_0, which represents the spatial scale of turbulence, is, at most, 1.5 meters at the best ground observatory sites; thus large infrared telescopes are still very much affected by seeing. The atmospheric dynamics above Poker Flats near Fairbanks, Alaska, have been extensively studied by analyzing the backscatter of the radar located there (Nastrom, Gage, & Balsley, 1982; Nastrom, Gage, & Ecklund, 1986). Based on these data, the mean r_0 in the near infrared (2.2 microns) at 12 km altitude is found to be about 10 meters. This indicates that the 6-meter aperture of POST will be diffraction-limited and that a simple tip-tilt correction (not adaptive optics) will be sufficient to compensate for all residual atmospheric effects. With an isoplanetic angle estimated at 5 arcminutes in radius, guide stars of magnitude 15 or brighter can be found over the entire sky with at least 95% probability.

2.3. Winds at altitude

Contrary to free-flying balloons, the aerostat is stationary and is subject to the full force of the wind at that altitude, a fact that significantly affects payload design and performance. The mean wind speed at 12 km altitude above Fairbanks is 17 m s^{-1} which is equivalent to 7 m s^{-1} (13 knots) at sea level. Exceptionally, wind speed can reach 40 m s^{-1} (equivalent to 32 knots at sea level).

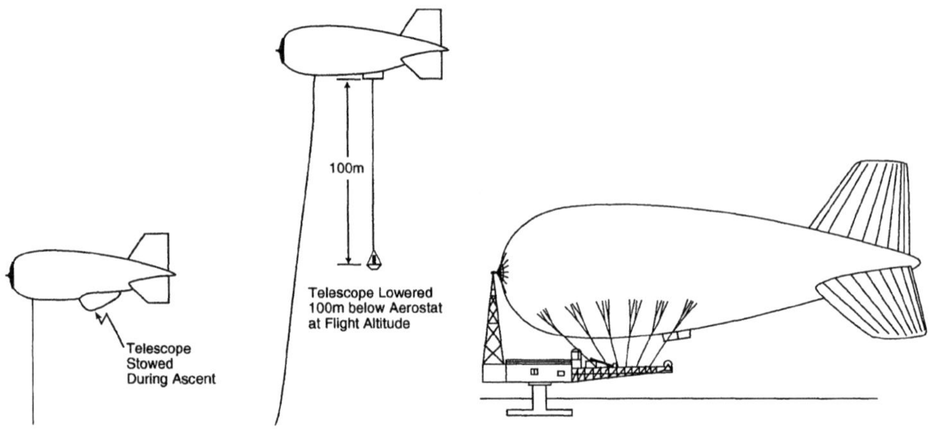

Fig. 2. Schematic view of the telescope and aerostat. The telescope is stowed inside the aerostat during ascent and lowered beneath the aerostat at flight altitude. The aerostat is shown on its mooring station at right.

3. The tethered aerostat

The aerostat consists of an aerodynamically shaped envelope filled with helium, with an internal chamber (ballonet) filled with air. In order to keep the envelope from becoming flabby due to thermal effects or wind, an automatic pressure system composed of sensors, valves and blowers keeps the internal pressure of the aerostat at about 50 mm of water pressure above that of the surrounding atmosphere by adjusting the amount of air in the ballonet. During ascent, expanding helium forces air from the ballonet into the atmosphere through automatic valves. During aerostat retrieval, air is automatically pumped back into the ballonet to compensate for the lowering internal pressure.

The tether maintains the aerostat in its position above the launch point. The ground facility is composed of the winching and mooring systems. While moored, the aerostat must be free to wind vane in order to minimize aerodynamic forces due to ground level winds. This is accomplished by attaching the tether hold-down pulley to a rotating arm cantilevered from the central winching and mooring station (Figure 2).

The telescope is suspended from the aerostat with a cable on a hoist. While on the ground and during ascent, the telescope is stored inside an enclosure in the aerostat for protection against contamination and bad weather. Upon reaching flight altitude, the enclosure is opened and the telescope is lowered to approximately 100 meters below the aerostat in order to reduce the amount of sky blocked near the zenith.

TABLE I
Aerostat characteristics

Maximum flight altitude	12 km
Overall length	90 m
Volume	32000 m^3
Aerostat mass	5700 kg
Tether	15 mm dia. Kevlar
Tether linear mass	0.15 kg m^{-1}
Tether mass at 12 km altitude	1800 kg
Power requirement on station	500 W*
Power requirement during descent	5 kW*
Retrieving and launching speed	150 m min^{-1}
Envelope lifetime	> 10 years

*Payload and ground winch excluded.

Tethered aerostat systems are very rugged and can be left flying for several weeks at a time. Typically, the aerostat is brought down for inspection and helium replenishment every two or three weeks, but downtime is minimal, on the order of a few hours total. The time required to winch the aerostat up or down is approximately one hour. Thanks to their sturdy and resistant envelope material, aerostats have long lifetimes, typically 10 years or more. The characteristics of a tethered aerostat capable of lifting a 1.5 ton payload to a 12 km altitude are shown in Table I.

The total power requirement is estimated at about 1 kW, 500 W each for the aerostat and the payload. Based on a 14-day mission, the total power consumption will be on the order of about 340 kW-hr. Power for low altitude aerostat-borne surveillance radar systems is fed via a cable imbedded in the tether. Because a power cable is not practical for POST's high altitude, we plan to generate power with fuel cells, a wind turbine, or by beaming microwaves from the ground.

Remote operation and telemetry of the aerostat and payload is accomplished by a microwave communication link with the local ground station.

4. The sparsely filled telescope

To take full advantage of the high resolution capability offered by the lower stratosphere in the near infrared, the telescope diameter should be as large as allowed by the residual seeing effects, which is about 6 meters. And, to take advantage of the low natural background, its optics should be cold, with

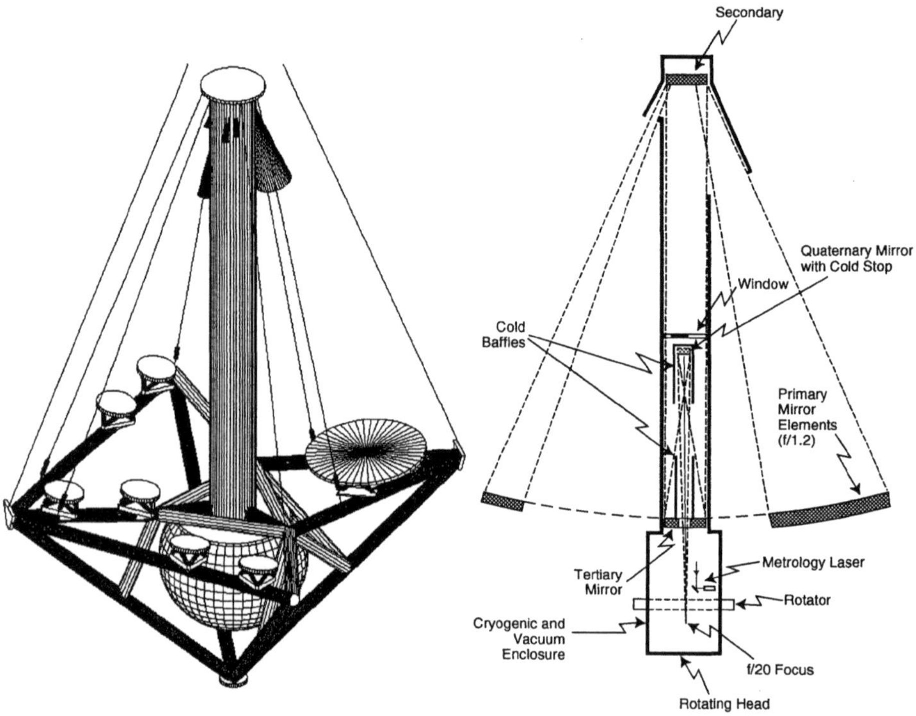

Fig. 3. Schematic view of the telescope. At left the overall view of the telescope, at right a cross section showing the optics, baffling and cold stop.

very low emissivity, and well baffled against internal and external radiation. On the other hand, the telescope should be light because of the aerostat lifting limitations, and with an open configuration to minimize wind induced disturbances.

With its diluted aperture design, and a mass of only 1300 kg thanks to state-of-the art active optics, pointing system and materials, POST, shown schematically in Figure 3, fulfills these conditions.

4.1. Aperture configuration

The sparsely filled aperture of POST has been optimized to provide excellent instantaneous coverage of the Fourier (uv) plane, with further improvement from Earth rotation. It is composed of one 1.8-meter diameter and six 0.6-meter diameter mirrors on a 6-meter outside diameter, leading to a total collecting area comparable to that of HST (4.2 m^2) and an aperture dilution factor of 15%. The positions of the subapertures were determined by following a method proposed by Cornwell (1988) which optimizes instantaneous uv plane coverage.

4.2. Optical configuration

Wind excitations, structure dynamics, and residual atmospheric seeing cause optical path differences that must be corrected to maintain diffraction limited performance. Because of their relatively large mass and inertia, the primary mirror elements cannot be moved very rapidly and this necessarily limits the bandwidth of the compensation that can be achieved. This compensation is better made on a smaller optical element further down in the system. The key to this approach is to locate this active element at a real image of the primary mirror surface; the corrections applied there will be fully equivalent to those needed at the primary mirror elements.

Such internal re-imaging requires more than the two surfaces of a traditional Cassegrain configuration; however, this is of little consequence because the additional surfaces are small enough to be cooled to reduce emissivity. The optical configuration selected for POST is composed of two stages. The first stage is a Mersenne 10-to-1 beam compressor which reduces the beam from the very fast f/1.2 primary to a sufficiently small size to permit light and cryogenically coolable optics. The resulting system is free of all aberrations to the third order, i.e., spherical aberration, coma, and astigmatism (Wilson, 1994; Schroeder, 1987). The second stage consists of a conventional aplanatic gregorian, delivering the desired final focal ratio. The final f/ratio is selected for critical sampling by the detector: f/20 for InSb arrays (30 μm pixels), or f/12 for HgCdTe arrays (18.5 μm pixels). The two stages are combined such as to form a real image of the primary mirror surface onto the small quaternary mirror which is used to cophase the overall system.

The central tube of the telescope serves as a baffle to prevent direct radiation or any single-scatter by Earth and celestial sources not in the telescope's field of view from reaching the detector. A second set of internal baffles and a cold stop placed at the exit pupil (quaternary mirror), all cryogenically cooled to about 165 K, prevent the surrounding "warm" parts of the telescope, including the areas in between the primary mirror elements, from radiating onto the detector. The primary mirror elements and the secondary are at the ambient temperature of $\sim$ 220 K. The tertiary and quaternary mirrors are enclosed in a controlled chamber to minimize dust contamination and cryogenically cooled to 165 K so as to render their emissivity negligible.

4.3. Cophasing of the primary mirror elements

This is the heart of the POST concept and the most challenging part of the project. Initial phasing of the primary mirror elements is obtained by an iterative process consisting of observing a bright star, determining the phase using phase retrieval, and correcting for the phase errors by adjusting the quaternary mirror sub-elements. Once the optimal position is obtained,

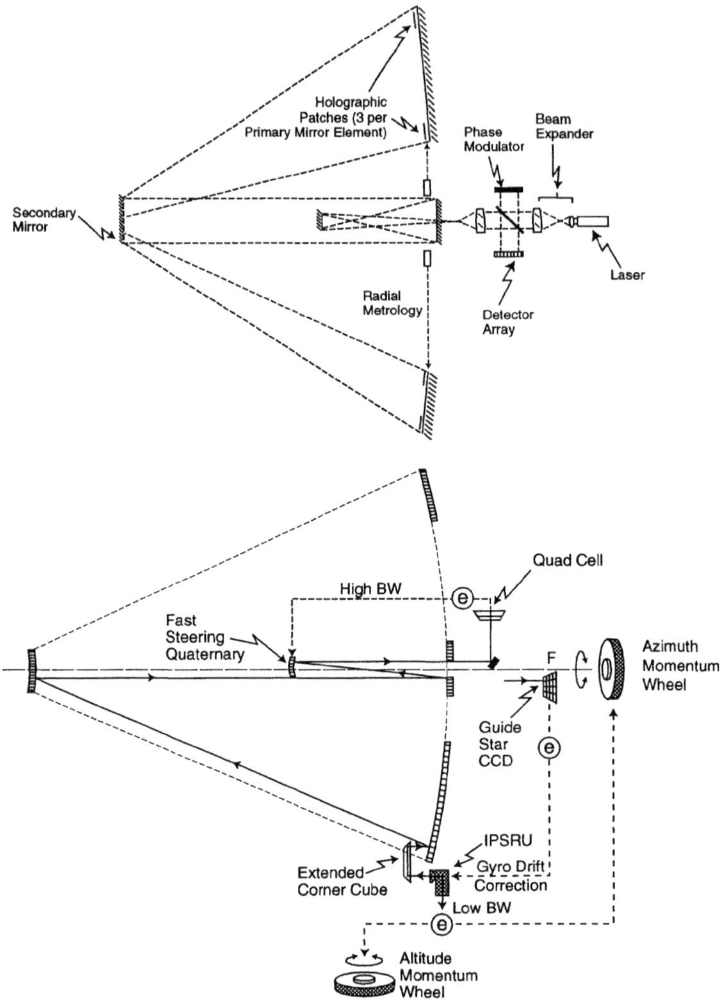

Fig. 4. Cophasing (top) and coaligning and guiding systems (bottom).

an internal metrology system is turned on to keep the optics phased. The metrology system is of the type invented by Bruning (1974) and developed by Kishner (1991). A laser located at the final focus illuminates the entire primary mirror surface via all intermediate mirrors (Figure 4). Holographic patches etched on each of the primary mirror elements (3 per element) return the laser beam back to the focal plane. After this roundtrip through the telescope optics, the return wave can be analyzed interferometrically to determine the optical path differences between each of the retro-reflectors. The system senses all internal optical path differences in the optical system, but not the path differences in the incoming wavefront. It is, of course, insensitive to an overall tilt of the line of sight, which must be corrected by a guiding system. Because it is also insensitive to a rotation of the primary mirror elements around the image of the laser error, a laser ranging system is used to measure the radial position of each primary mirror element. The corresponding value is fed into the internal system to correct for the error using the active quaternary mirror.

4.4. Pointing and guiding system

In order not to degrade the imaging quality of the optics, the line of sight should be stable to a fraction of the angular resolution, typically 10% of it. In the near infrared at 2.2 microns, where the full width half maximum image size is about 80 milliarcseconds, the line of sight should thus be stable to about 8 milliarcseconds. This is a stringent requirement, comparable to that of the Hubble Space Telescope, which is especially difficult to meet since POST is faced with specific disturbances due to gravity and wind which are not encountered in space. It is clear that the "body pointing" method, in which the entire body of the telescope is accurately pointed, would not be adequate. The solution is to complement the overall telescope pointing by adjustment of an optical element with low inertia and fast response time, for which the small and active quaternary mirror is a natural choice.

Two reaction wheels, one each for the altitude and azimuth axes, are used to point the telescope to about one arcminute accuracy. Wheel speed saturation is prevented by the use of flaps to cancel aerodynamic torques. The quaternary is then used as a fast steering mirror for adjusting the line of sight at the few milliarcseconds level.

Sensing the pointing errors is itself done in two layers: the high frequencies of the line of sight excursion (above a few Hz) are sensed by a fast response inertial system, while an absolute sensing system measures the slow drift of the inertial system using a guide star in the field of view. The inertial sensing makes use of a novel device which does not simply sense the inertial motion of the telescope mechanical boresight, as in traditional space pointing systems, but directly the *inertial direction of the line of sight itself.* This device, developed by Draper Laboratories is composed of an inertially stabilized platform supporting a laser which feeds an alignment beam into the telescope (Gilmore, 1992). Called "IPSRU", which stands for Inertial Pseudo Star Reference Unit, the system in effect creates a pseudo star which is very bright and can thus be used to control the line of sight at a very fast rate. IPSRU's inertial sensing uses gyroscopes which will drift eventually, but with a time scale long enough to be compensated by the field star guiding system. POST implementation of this system is schematically shown in Figure 4. The laser beam reflects off all mirrors in the train like authentic starlight and the resulting spot in the focal plane is sensed by a quadcell. The measured spot excursions are used to correct the line of sight by steering the quaternary mirror. With a 5 mW laser, the spot centroid can be measured at more than 100 Hz rate to a few milliarcseconds accuracy and allows for the correction of vibration in the optical train induced by wind or mechanical excitation.

5. Predicted telescope performance

Although the POST concept lends itself to a variety of observations from the visible to longer infrared, it is in the near infrared, between 2 and 10 microns, and especially from 2 to 3 microns that its advantages are the most significant.

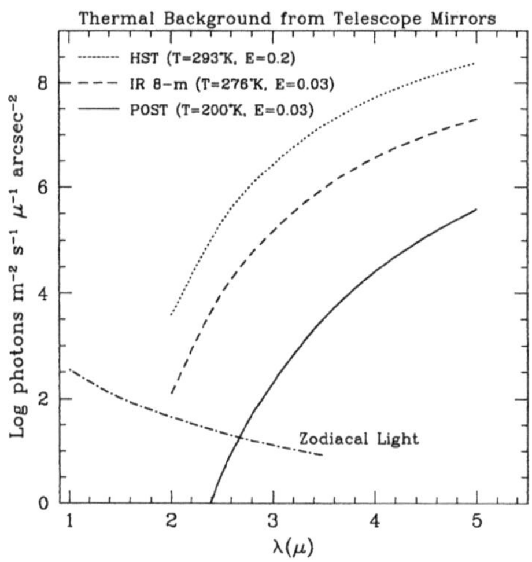

Fig. 5. Thermal background from the mirrors for HST, an IR-optimized 8-m ground telescope, and from POST. The background from scattered zodiacal light is at an ecliptic latitude of 45 degrees.

At 2.6 microns the thermal emission from the primary and secondary mirrors which are at the ambient temperature of 220 K and silver coated (0.03 emissivity) is equal to the scattered light from zodiacal dust at an ecliptic latitude of 45 degrees (Figure 5). The background at this wavelength is approximately 50 times fainter than the darkest part of the optical spectrum, which occurs near 500 nm. Longward of 3 microns the background begins to rise again because of thermal emission from the zodiacal dust. Consequently, the 2 to 3 micron window is the darkest part of the spectrum from optical wavelengths to 100 microns. The combination of high resolution images and a very dark sky make this the spectral region of choice for observing the redshifted light from galaxies and clusters of galaxies at moderate to high redshifts.

We have compared the performance of POST at infrared wavelengths to the HST and to an infrared-optimized 8-m telescope on Mauna Kea.

We used model atmospheres to calculate the atmospheric transmission and emissivity for Mauna Kea and POST. The 8-m was assumed to be nearly diffraction-limited, with a Strehl ratio of 0.5. Both POST and the ground-based telescope were assumed to have emissivities of 0.03 for their mirrors, with the POST optics being at 220 K and the ground-based telescope at 276 K. HST which is not infrared optimized has a total optical system emissivity at about 0.2 and its optics are at 276 K. The results are shown in Figure 6. From 2 to 8 microns POST will be superior to HST or an infrared-optimized 8-m on Mauna Kea.

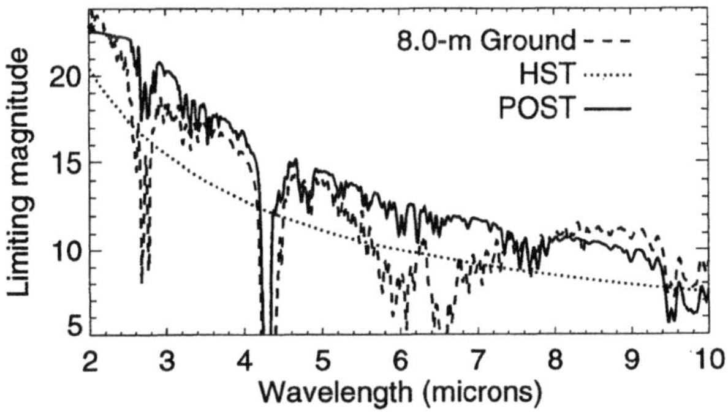

Fig. 6. Estimated limiting magnitudes in the infrared for POST, *HST*, and an infrared-optimized 8-m telescope on Mauna Kea. A point source of the given magnitude is detected with SNR=10 in a 1 hr exposure for a 0.5% wide filter bandpass.

6. Science with POST

POST will provide a unique capability for seeing spatial detail in faint extended sources at near-infrared wavelengths between 2 and 7 microns. At these wavelengths POST will detect stars and galaxies which are two or more magnitudes fainter than can be detected by any present or planned telescope on the ground or in space. We plan to use POST's unprecedented near-infrared spatial resolution and sensitivity to attack a suite of outstanding problems ranging from planetary science to cosmology. We will use POST to search for planetary systems, protoplanetary disks, and brown dwarfs around nearby stars. POST will be especially suitable for observing the formation of galaxies and clusters of galaxies in the early universe and following their evolution from then to now. By observing Type Ia supernovae in high redshift clusters, we can measure the curvature of the Universe and thereby determine if the expansion of the Universe will either continue indefinitely or eventually slow and stop, and then begin to fall back in upon itself.

Acknowledgements

We would like to thank Herbert Scharbeck and Robert Ashford of the TCOM corporation for the definition of the aerostat platform, Jay Tilly of the Ball Aerospace Corporation for discussions concerning the telescope concept, and Robert Williams and Peter Stockman of the Space Telescope Science Institute for their support and encouragement.

References

Bruning, J.H., et al.: 1974, 'Digital wavefront measuring interferometer for testing optical surfaces and lenses', *Applied Optics* **13, 11**, 2693

Cornwell, T.J.: 1988, 'A Novel Principle for Optimization of the Instantaneous Fourier Plane Coverage of Correlation Arrays', *IEEE Tr.* **AP-8**, 1165

Gilmore, J., et al.: 1992, 'Inertial Pseudo-Star Reference Unit', *IEEE Proc.* **PLANS Symposium.**,

Kishner, S.J.: 1991, 'High bandwidth alignment sensing in active optical systems', *Proc. SPIE, Analysis of Optical Structures* **1532**, 215

Nastrom, G.D., Gage, K.S., Balsley, B.B.: 1982, 'Variability of C_n^2 at Poker Flat, Alaska, from mesosphere, stratosphere, troposphere (MST) Doppler radar observations', *Opt. Eng.* **21**, 347

Nastrom, G.D., Gage, K.S., Ecklund, W.L.: 1986, 'Variability of turbulence, 4-20 km in Colorado and Alaska from MST radar observations', *J. Geophys. Res.* **91**, 6722

Schroeder, D.J.: 1987, *Astronomical Optics*, (San Diego, CA: Academic Press)5

Wilson, R.N.: 1994, 'Karl Schwarzschild and telescope optics', *Rev. Mod. Astron.* **7**, in press

INFRARED IMAGING SURVEYOR: IRIS

T. MATSUMOTO
Department of Astrophysics, Nagoya University Chikusa-ku, Nagoya, Japan

Abstract. Based on the experience of the IRTS, a new infrared satellite mission, Infrared Imaging Surveyor:IRIS is now being proposed to ISAS. The IRIS will be thrown into the sun synchronous orbit at 900km altitude by the MV rocket which is now under development in ISAS. With use of new cryogenic technologies, considerably light weight with large aperture telescope is designed. Owing to the large format states-of-art array detectors, the IRIS has an enormous capability for the survey observation, and will provide new discoveries on the formation and evolution of galaxies, brown dwarfs, protoplanetary disk, interstellar matter and so on.

1. Introduction

In Japan, a first infrared satellite mission, IRTS is now going on (Murakami et al., 1994). The IRTS, however, is a mission for specific targets, and the observation period is not long enough. Based on the experience obtained in developing the IRTS, a new satellite mission has been examined in Japan. Fortunately, ISAS is now developing a new launching rocket, MV, which has the capability to launch a 2 ton satellite into the LOE and 1 ton satellite into the sun synchronous orbit.

The MV has a considerably higher launching capability than the previous launcher, however, is not so powerful as to use the ISO orbit (high eccentric orbit) or a solar orbit proposed for a new SIRTF. Under the given condition, we examined the best infrared mission and constructed the mission plan of IRIS: Infrared Imaging Surveyor.

The IRIS will be thrown into the sun synchronous orbit to launch as large aperture telescope as possible. The IRIS is not the observatory type mission but is optimized for the survey observation. With use of the large format state-of-art array detectors, significantly deep survey for wide sky coverage will be achieved.

In order to reduce the weight and to extend the life of the coolant, new techniques are introduced. One is a mechanical cryo-cooler to cool the vapor cooling shield, and another is an effective application of the radiative cooling. The IRIS will be the first mission to apply these new technologies, which will provide valuable experiences for the future space infrared mission.

The IRIS mission is now being proposed to ISAS and will be evaluated in 1994. If the IRIS is approved, it will be launched in 2001 after the next X-ray mission ASTRO-E.

Space Science Reviews **74**: 113–117, 1995.

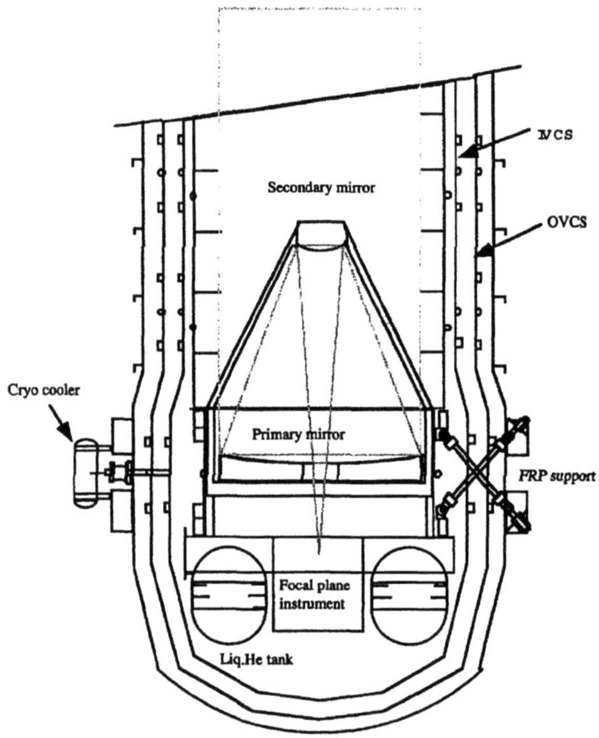

Fig. 1. Conceptual design of the IRIS cryostat.

2. The IRIS mission

Due to the serious restriction on the weight, a sun synchronous orbit at 900 km altitude is adopted. The IRIS flies along the twilight zone which makes easy to avoid the direct incidence of the sun light. The basic attitude control is to rotate the IRIS uniformly once in a orbital motion so that the optical axis always direct the zenith. Under this basic control, two observation modes are currently considered. One is a general survey as the IRAS did, which will be used for the survey in far infrared and submillimeter wavelengths. Another is pointing observation. Once or twice in one orbital revolution, the satellite is fixed in the inertia space for about 10 min during which imaging observations in near and mid infrared wavelengths will be attained.

Since IRIS flies along the twilight zone, the bottom surface of the bus module always directs towards the earth surface and one side always faces the sun. This allows a simple thermal design, giving < 150 K for the temperature of the outer wall of the cryostat.

TABLE I
Characteristics of the focal plane instruments of the IRIS

Infrared Camera	
FOV	$20' \times 20'$
Near infrared band	1024×1024 InSb array
	1.2 arcsec/pixel
	2-5 μm, wide band imaging, grism spectroscopy (option)
Mid infrared band	256×256 SiAs(or SiSb) BIB array
	4.8 arcsec/pixel
	5-30 μm, wide band imaging

Far Infrared Scanner	
50 μm band	two 1-dim. GeGa arrays (20×2)
	30 arcsec/pixel
	FOV $10' \times 30'' \times 2$
90 μm band	two 1-dim. GeGa arrays (20×2)
	30 arcsec/pixel
	FOV $10' \times 30'' \times 2$
160 μm band	two 1-dim. GeGa (stressed) arrays (12×2)
	45 arcsec/pixel
	FOV $10' \times 45'' \times 2$
submillimeter bands (optional)	bolometers

Figure 1 shows a conceptual design of the IRIS cryostat. The cryostat is a vacuum vessel in which 200 liter of liq. He is contained. The telescope is thermally isolated from the liq. He tank and is cooled by the evaporated He gas. Two vapor cooling shields are installed. Outer vapor cooling shield (OVCS) has a radiative cooler that is opened to the blank sky, while the inner one (IVCS) is cooled by 2-stage Stirling cooler. The engineering model of the 2-stage Stirling cooler already achieved requirement of 200 mW absorbing power at 20 K (Kyoya et al., 1993). The space qualified system is under development in Sumitomo Heavy Industry.

Thermal calculation of the cryostat indicates that the temperature of the telescope is $\simeq 5$ K which is low enough for the far infrared observation. The total heat load on the liq. He tank is 15.3 mW assuming 10 mW power dissipation for the focal plane instruments. The holding time of the liq. He is about 400 days. After liq. He is exhausted, temperature of the telescope could be around 20 K which is low enough to continue the near infrared observation.

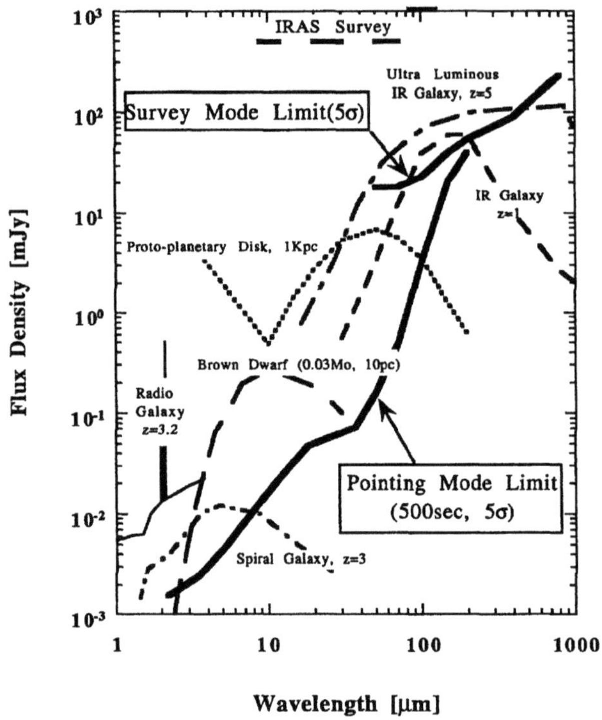

Fig. 2. Detection limits of the IRIS and flux levels of typical targets.

3. Telescope and focal plane instruments

The aperture size of the telescope strongly depends on the available weight for the satellite, however, 80 cm aperture is currently assumed. The telescope is a Ritchey-Chretien type with focal ratio of F/6. Image quality of 2 arcsec is expected.

At the focal plane, two instruments are under consideration as is summarized in Table I. The Infrared Camera covers near and mid infrared regions, and is used for pointing observations. Large format arrays, which cover 20 arcmin field, are installed. The Far Infrared Scanner consists of six 1-dimensional GeGa arrays, which correspond to 50, 90, 160 mm, respectively. Submillimeter bands with bolometers are considered as an optional instrument.

4. Astronomy with the IRIS

In Figure 2, detection limits of IRIS are compared with several observation targets. In near and mid infrared regions, integration time of 500 s is assumed.

Detection limits in near infrared region are mainly due to detector noise, while natural background noise is dominant in mid infrared region. Confusion noise becomes significant at wavelengths longer than 100 μm even in the case of the general survey.

Since the IRIS is a survey mission, various observation targets are expected. With the deep survey for limited sky (near and mid infrared regions) and the shallow general survey (far infrared region), following scientific results will be expected.

- Birth and evolution of the galaxies
  - A search for the primeval galaxies
  - Trace the evolution of the normal galaxies
  - Origin and evolution of the infrared galaxies and AGN
- Star formation and interstellar matter
  - A search for the brown dwarfs
  - IMF towards the low mass end
  - Interstellar dust and gas
- Extra-solar system
  - Evolution of the dust cloud around the T-Tauri stars
  - Vega-like stars, dust shell around the main sequence stars

References

Murakami, H. et al.: 1994, *Astrophys. J.* **428**, 354.
Kyoya, M. et al.: 1993, *Cryogenics* **35**, 431.

OPTICS OF THE FAR INFRA-RED AND SUBMILLIMETRE SPACE TELESCOPE

RICHARD HILLS
MRAO, Cavendish Laboratory, Cambridge, U.K.

Abstract. Some of the considerations in the design of the telescope for FIRST are discussed. It is pointed out that instruments that operate at submillimetre wavelengths need to be analyzed with techniques derived from both the Radio and Optical/IR traditions. The issue of emissivity of reflector surfaces is also discussed.

Key words: antennas – far-IR – submillimetre

1. Introduction

The European Space Agency's FIRST project is aimed at providing a large diameter telescope in space operating in the wavelength range 100 to 600 microns, to be flown early in the next century. For this a lightweight but rather accurate main reflector is required. It will be passively cooled but the temperature is unlikely to be lower than about 150 K. Although this inevitably means that the detectors will see some background emission from the telescope, careful design should keep this to quite low levels.

There is a very large body of knowledge relating to the design of antennas for centimetre and millimetre wavelengths, and of course a great deal is known about the design of optical and infra-red telescopes. Although there are many parallels between the approaches taken in these two wavebands, there are also major differences. Some of these are simply questions of terminology – 'secondary mirror' versus 'sub-reflector' – but many are quite basic and arise from the great difference in the scale of the optical components when measured in terms of the wavelength at which they operate. The far-infrared and sub-millimetre wavebands lie at the point in the spectrum where these approaches meet: it is important to learn from both of them and it is also essential to check the validity of approximations and assumptions.

2. Basic Layout

A range of possible schemes for the basic layout were considered before selecting the baseline design. Off-axis Cassegrain and Gregorian telescopes have the advantage of providing a very "clean" optical path: this gives low background, low sidelobes, reduced scattering ("stray light") and essentially zero standing wave ratio (see below). The manufacture of the optics is however considerably more difficult than for a symmetric system and, for a large antenna like that planned for FIRST, the positioning of the optics would have to take place in orbit, which is a high risk procedure.

Space Science Reviews **74**: 119–123, 1995.

A system with an elliptical primary was also discussed as a way of increasing the collecting area of the telescope while still fitting it inside the shroud of the launcher. This is however not very attractive because of the elliptical shape of the diffraction-limited beam. It is at present not relevant anyway because cost limitations have enforced descoping of the project to the point where a circular primary will fit inside the shroud.

The baseline for the geometry is therefore classical Cassegrain optics, with the possibility of applying some shaping to both surfaces to improve the image quality over the whole field of view, which at 15 arc minutes is quite large. The focal ratio of the primary is likely to be around $f/D = 0.5$ – very short compared to typical optical and IR designs but a little longer than most radio antennas, which are usually in the range 0.3 to 0.4. This design has good structural properties and the symmetric optics are relatively easy to manufacture, making the achievement of the necessary surface accuracy secure.

3. Problems of On-axis Cassegrain Antennas

Some parts of the optical path from the source to the detectors are inevitably blocked in an on-axis design. This will produce sidelobes in the diffraction pattern of the telescope, but provided the secondary mirror is small compared to the primary and the supports for it are thin, these are not a major problem. More serious difficulties could arise from the signals that the detectors may pick up from these blocked regions. On optical telescopes (and cryogenically cooled ones like ISO and SIRTF) such areas are usually covered by "black" material to absorb any light that is scattered into these regions. Because the FIRST optics will be relatively warm and the detectors will be operating in the Rayleigh-Jeans part of the spectrum, where any absorber is also an emitter, this approach cannot be used: the emission would increase the background seen by the detectors to an unacceptable degree. Instead we will need to cover structures like the supports for the secondary mirror with highly reflecting material and shape these covers in such a way that the detectors only "see" cold sky reflected in them. That this approach is acceptable is a result of the fact that there are relatively few very strong sources of emission in the sky at far-infrared and submillimetre wavelengths, in contrast to the optical situation.

The relatively long operating wavelength means that diffraction has to be taken into account in designing the optics. For example, diffraction by the edge of the secondary mirror inevitably means that the detectors will pick up signals from outside the primary. Some form of highly reflective "collar" around the primary mirror will be needed to make sure that this region again appears cold. (The alternative would be to use an undersized secondary mirror, but this would of course reduce the effective aperture of the telescope significantly .)

Another problem which will be unfamiliar to optical designers is that of standing waves. These arise when some part of the signal (either signals from the astronomical source or perhaps noise arising inside the instruments) is reflected from the detector, travels out onto some part of the telescope (for instance the centre of the secondary mirror) and is reflected back into the detector again. Interference between the original signal and the reflected one then produces a "ripple" on the spectrum observed. The period of the ripple, measured as the the spacing in frequency between the peaks, is given by the speed of light divided by the total path length travelled by the interfering signal. This spacing turns out to be similar to the width of the Doppler-broadened spectral lines which FIRST will often be observing. We must therefore take steps to prevent reflections – for instance by placing a small cone at the centre of the secondary mirror – but again a full diffraction analysis must be employed in the design.

4. Mirror Fabrication

Early work was directed towards a telescope with an 8–metre aperture. This would have required a mirror made of segments which would be deployed in orbit. To achieve the required accuracy would have been difficult and it is now clear that the development of such a system is too expensive for this project.

For the smaller diameters which were subsequently considered, 4.5 and 3 metres, a range of techniques are available. The approach which follows most closely that used on large ground-based millimetre-wave telescopes is to make the surface out of a number of independent panels mounted on a stiff supporting structure. This has the advantage of separating the problem of making a reflecting surface which is locally smooth, from that of providing a stiff and stable support which maintains the overall figure of the mirror. Disadvantages are the emission and scattering from the gaps between the panels and the danger that the relative alignment will be lost during launch.

One alternative that would provide a continuous surface is a single shell, probably made of carbon-fibre reinforced plastic (CFRP) coated with a metal film to give high reflectivity. This would be supported from a stiff backing structure by a large number of pins. The pins would be adjusted to bring the surface into the correct shape and then epoxied firmly into place. More recently the "single panel" concept has been developed. This uses two CFRP skins separated by a honeycomb structure to provide a lightweight primary reflector with excellent structural integrity and low thermal deformations.

It is likely that there will be some deformations of this mirror as a result of loss of water and cooling to 150 K. These can however be measured in a simulated space environment and if they take the surface errors outside the permitted limits of about 5 microns rms then it will be possible to correct them by constructing a secondary mirror with equal but opposite errors. This correction can only work perfectly at one point in the field of view and at short wavelengths where ray optics are valid. The deformations will however have large spatial scales and as a result the cancellation of the errors will in fact be very good over the whole field of view and the diffraction effects will not be important.

5. Background Emission

Some of the instruments on FIRST will be high-resolution spectrometers using either heterodyne or photoconducting detectors. For these instruments the noise arising in the detectors themselves is likely to be dominant and the additional power arriving from the relatively warm optics will have little effect. However important aspects of the scientific programme require the use of relatively broadband detectors and these are likely to be operating in the background-limited regime. It is therefore important to reduce the emission from the reflecting surfaces as far as is practical.

The first point to note is that in the Rayleigh-Jeans part of the spectrum the power emitted is proportional to temperature times the emissivity, ϵ. We should therefore pay equal attention to reducing each of these.

The best way of obtaining a high-reflectivity surface over a wide range of wavelengths is to use a film made from a highly conducting metal. In the submillimetre and far-IR wavebands most metals still have bulk conductivities close to their d.c. values. The best conductors are silver and copper, with gold and aluminium having about 50% higher losses. With good conductors the reflection actually takes place in a very thin layer close to the front surface of the film: the currents decay exponentially beneath the surface over a distance characterised by the skin depth, which is about 100 nm at a wavelength of 1 millimetre and 30 nm at 100 microns. The film does not therefore need to be more than about 0.5 microns thick but it must be smooth and continuous on scales down to the skin depth and somewhat below. The condition of the metal must be such that its conductivity is close to the bulk value. In this case the effective surface impedance is given by the bulk resistivity divided by the skin depth, giving values of order 0.1 to 0.5 Ω. The reflectivity can then be calculated from the mismatch between this (complex) impedance and that of free space (377 Ω). This leads to a convenient expression for the theoretical emissivity of $\epsilon = 0.002\sqrt{\rho' f'}$, where ρ' is the resistivity measured in units of $10^{-8}\,\Omega\,\mathrm{m}$ and f' is the frequency in THz. At room temperature ρ' is about 1.6 for silver and 2.6 for aluminium so that emissivities in the region of 0.2 to 0.6% are to be expected in the part of the spectrum where FIRST will operate. My understanding is that laboratory measurements of good quality metal films are consistent with these figures.

It is worth noting that the electrical resistivity of metals falls substantially at low temperatures, being roughly half the room temperature value at 150 K. This means that the background power seen by the detectors will scale with roughly the three-halves power of temperature. As a result of these frequency- and temperature-dependent factors the overall emissivity of the FIRST telescope should be much lower than might be expected on the basis of optical and near-IR experience.

Clearly contamination of the surface will cause increased emissivity. This could be a serious problem especially if the mirror surfaces are relatively cold and other parts of the spacecraft which are likely to contain volatile materials, such as the heatshields, are warm. However the loss of reflectivity due to thin films of lossy material at the wavelengths we are considering here is not a great as one might expect from experience in the optical regime. The reason for this is that the tangential electric field near the surface of a good conductor is extremely low. This means that very little energy is dissipated in the contaminating layer, provided its thickness is much less than a wavelength. (Essentially the lossy material finds itself at a null in the standing-wave pattern.) For similar reasons, small particles of dust, etc, are not likely to cause significant loss of signal or increase in background noise. Note also that it should be possible to employ coatings on top of the reflecting surface should this be necessary to prevent attack before launch or in orbit.

6. Conclusions

At far-IR and sub-millimetre wavelengths large apertures are essential for achieving good angular resolution, which is the key to obtaining a good understanding of complex astronomical objects and to avoiding problems of confusion when searching for faint objects. To keep costs reasonable, large mirrors have to be relatively warm. Because the emissivity of mirrors can be kept low at these wavelengths, the increase in the background seen by the detectors is, however, not as great as is often supposed.

Acknowledgements

Much of the work outlined here has been undertaken by those managing the FIRST studies at ESTEC, their contractors and the members of the Telescope Working Group, J.W.M.Baars and J-M Lamarre. It is pleasure to thank them all for their contributions.

SIRTF - THE MODERATE MISSION

M. W. WERNER and L. L. SIMMONS
Jet Propulsion Laboratory, California Institute of Technology, Pasadena, CA 91109

Abstract. The Space Infrared Telescope Facility (SIRTF) has been planned by NASA and the US scientific and aerospace communities as a cryogenically-cooled observatory for infrared astronomy from space. Within the past few years, severe pressures on NASA's budget have led to the cancellation of many programs and to dramatic rescoping of others; SIRTF is in the latter category. This paper describes the resulting redefinition of SIRTF and the technical innovations which have made it possible to package SIRTF's key scientific capabilities into the envelope of a moderate-class mission.

1. Background and Introduction

The Space Infrared Telescope Facility (SIRTF) has been planned by NASA and the US scientific and aerospace communities as a cryogenically-cooled observatory for infrared astronomy from space. SIRTF will build on the scientific and technical accomplishments of the predecessor IRAS, COBE, and ISO missions. However, SIRTF will take a major step beyond these missions by combining - for the first time - the intrinsic sensitivity of a cryogenically-cooled telescope for infrared astronomy from space with the great imaging and spectroscopic power of large format, low-noise infrared arrays. The scientific capabilities of this combination are very powerful, and the many exciting results from IRAS and COBE have whetted our appetite for a more penetrating look at the infrared sky. As a result, in 1991, SIRTF was designated by the National Academy of Sciences "decade review" of astronomy and astrophysics - chaired by John Bahcall - as the highest priority new major mission for all of US astronomy in the 1990's.

2. Constraints and requirements

The chief constraint which has driven the redefinition of SIRTF is the requirement that the development phase cost be no greater than \$500 M in real year dollars. The system described below has in fact been even more tightly constrained by a self-imposed requirement of \$400 M real year dollars, including contingency. In response to this constraint, the SIRTF Science Working Group has rebuilt SIRTF's scientific and instrument requirements around four major scientific themes which parallel the scientific priorities identified for SIRTF in the Bahcall report. These themes are:

1. Protoplanetary and Planetary Debris Disks
2. Brown Dwarfs and Superplanets
3. Ultraluminous Galaxies and Active Galactic Nuclei
4. The Early Universe

Space Science Reviews **74**: 125–138, 1995.

Only facility and instrument capabilities essential for the study of these four problems are allowed to drive the cost and complexity of SIRTF. However, because a system optimized for the exploration of these questions will have powerful capabilities for the study of many other problems, SIRTF retains it broad scientific capabilities and appeal.

The SIRTF scientific and engineering teams have adopted an initial set of facility and instrument requirements to guide the system design process. These requirements, which are consistent with our four scientific goals, are:

1. Aperture - 85 cm
2. Wavelength Range
   - Imaging - 3-180 μm
   - Spectroscopy - 4-40 μm
3. Lifetime - 2.5 yrs.
4. Image Quality - 50% encircled energy within 2″ diameter at 3.5 μm
5. Optics Temperature - 5.5 K
6. Instrument Accommodations:
   - Cold mass - 50 kg
   - Cold power dissipation - 10 mW
   - Cold instrument volume - 0.2 m^3

In addition, a new "warm-launch" system architecture was chosen for investigation. In this approach, the telescope is launched at ambient temperature and is cooled on orbit by a combination of passive radiation and the boil-off helium gas from the cryogen tank. It therefore differs from the "cold-launch" architecture used by IRAS and ISO, in which the telescope is contained within an annular cryogen tank. The features of SIRTF's warm launch architecture are discussed further below. A very thorough discussion of purely radiatively-cooled space telescopes is given by Hawarden et al. (1992). In areas where the analyses overlap, such as the prediction of outer shell temperatures, the results of the SIRTF calculations agree with those presented by Hawarden et al. (1992).

3. System Description

3.1. Orbit and Launch Vehicle

An important feature of the new SIRTF mission is the adoption of a solar orbit (Figure 1). To reach this orbit, the spacecraft is launched with slightly greater than the terrestrial escape velocity in such a manner that it ends up trailing the Earth in its orbit around the sun. This orbit makes better use of launch capability than does a conventional Earth orbit, and it permits excellent, uninterrupted viewing of a large portion of the sky without the need for Earth-avoidance maneuvers. In addition, the absence of heat input from the Earth provides a stable thermal environment and allows the exterior of the telescope to reach a low temperature via radiative cooling.

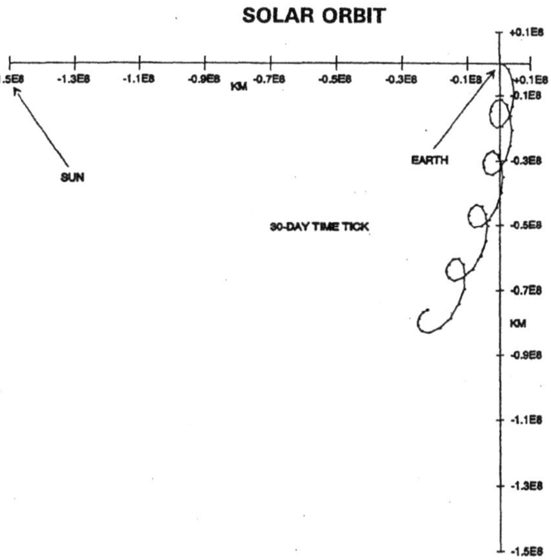

Fig. 1. SIRTF's solar orbit is shown in a coordinate system fixed in the Earth-Sun rotating frame. SIRTF will follow the Earth around the Sun, reaching a distance of about 0.6 au over the five-year interval shown in the figure. The annual loops occur because the eccentricity of SIRTF's orbit would be slightly different from that of the Earth's orbit.

A $\simeq$1 m diameter transmitting antenna fixed to the bottom of the spacecraft is used twice per day to downlink 12 hours' stored science data to stations of NASA's Deep Space Network. In this manner, an adequate average data rate (45 kbps) can be maintained over the lifetime of the mission with only a minimal loss of efficiency.

Current plans call for the use of a Delta launch vehicle for SIRTF. The Delta 7920 will lift about 700 kg to SIRTF's solar orbit. If this capability provides inadequate mass margin, it is possible to add an additional kick stage to the payload and increase the capacity to solar orbit to upwards of 1000 kg.

3.2. Thermal and Cryogenic Design

Figure 2 is a sketch of the warm-launch system architecture. In this warm launch system, the telescope at launch is in good thermal contact with the outer case of the liquid helium dewar and with the outer shell which surrounds the telescope. The instruments are within the dewar and are at < 2 K at launch and throughout the mission. The shell-telescope-dewar combination is at ambient temperature when launched but cools rapidly by radiation to 50-70 K within a few days on orbit (Figure 3). At that point, the heat switch (mechanical or gaseous) which provides the thermal contact between

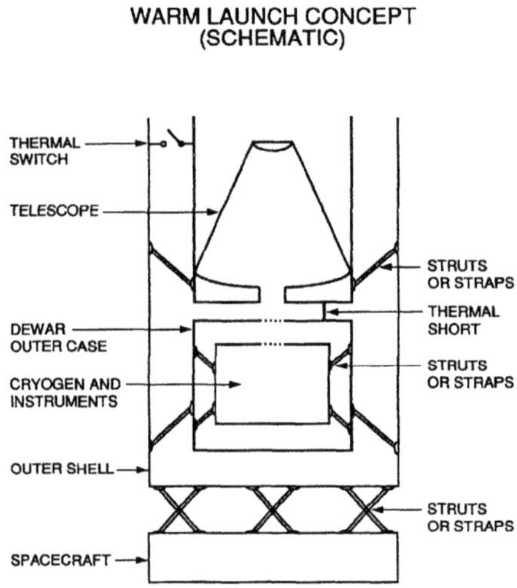

Fig. 2. A schematic representation of the essential features of the warm-launch architecture adopted for SIRTF. The dashed lines in the dewar outer case and at the top of the cryogen chamber represent windows which will open on-orbit to admit infrared radiation into the instrument volume. See text for a detailed discussion.

the shell and the telescope-dewar assembly is opened. The shell remains at the radiative equilibrium temperature (50-70 K), while the telescope and the dewar case are vapor-cooled by the boil-off cryogen, and, to a lesser extent, by radiation through the telescope aperture. Calculations indicate that cooldown of the telescope to its final operating temperature of 5.5 K may take 40-60 days, and the 10 mW of focal plane power dissipation provides adequate vapor cooling to maintain it at this temperature throughout the mission.

However, an optics temperature of 5.5 K is required only for observations at SIRTF's longest wavelengths, 100-180 μm. Observations at wavelengths shortward of 10 μm, for example, require only telescope temperatures below 50 K, so that on-orbit checkout and short wavelength scientific observations could commence just a few days after launch. The current sizing of the cryogen system calls for a helium tank with a capacity of 330 liters. This provides adequate cooling capacity for the initial cooldown and a subsequent on-orbit lifetime of 2.5 years at the average power dissipation of 10 mW.

This top-level description of the warm-launch configuration points up some of the interesting technical challenges inherent in this approach. These range from identifying the high emissivity, high conductivity materials required to optimize the initial on-orbit cooldown to designing a dewar which operates efficiently both with an ambient temperature outer case (on the ground) and with a very cold outer case (on orbit). These challenges must

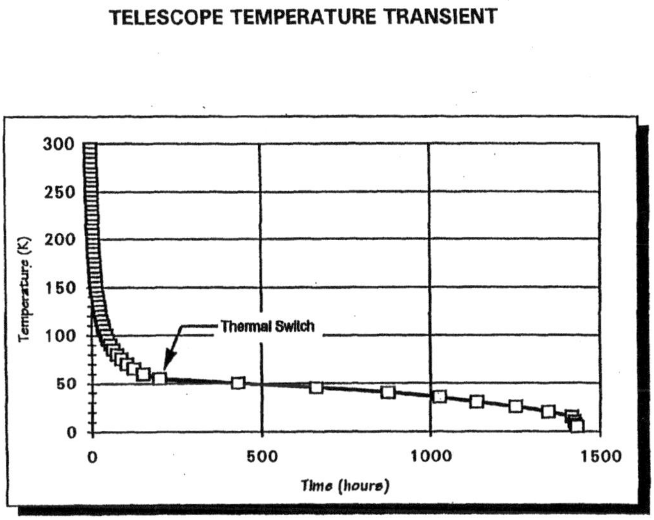

Fig. 3. A model calculation of the thermal behavior of a system such as that shown in Figure 2. Radiative cooling brings the telescope temperature down rapidly to about 50 K in the first ≃250 hours after launch. At this time, the thermal switch opens and the telescope decouples from the radiating outer shields. The subsequent cooling to the final operating temperature of 5.5 K is produced by the enthalpy of the vapor escaping from the Helium tank.

be weighed against the potential benefits of the warm-launch approach. Chief among these are: (1) Placing the telescope outside of the cryogen tank permits the aperture of the telescope to be increased without increasing the size (and mass) of the dewar, which should allow a larger telescope to be launched on a given launch vehicle; (2) Because the telescope is launched warm, a cryogenic shake test of the telescope is no longer required (it appears that there are other potential savings and simplifications in test and integration as well); (3) The warm-launch architecture is a step towards the future, because it can readily be adapted to telescope apertures larger than the ≃1 m size of SIRTF and its cold launch predecessors, ISO and IRAS.

3.3. Optics and Optics Technology

The other facility subsystem which is receiving particular attention at this phase is the telescope, including the primary and secondary mirrors and associated support and mounting structures. The telescope design is driven by two competing requirements: (1) Low mass - needed both for rapid on-orbit cooling and for compatibility with a Delta launch; and (2) Good optical performance at cryogenic temperatures - our fundamental requirement of 50% encircled energy within 2″ corresponds roughly to an RMS wavefront error of 0.1 waves at 3.5 μm when the effects of the telescope's central obscuration are taken into consideration.

Examination of existing materials and experience led to the conclusion that the parameters of the SIRTF telescope were beyond what had been convincingly demonstrated in previous cryogenic systems. As a result, a Telescope Technology Testbed program has been initiated, primarily under the sponsorship of NASA's Office of Advanced Concepts and Technology. The objective of this program is the design, fabrication, and test - by the Fall of 1995 - of a telescope which is consistent with SIRTF's mass and image quality requirements. The mass of the telescope is to be about 40 kg, and the optical quality at 5.5 K shall correspond to a wavefront error of 0.1 to 0.13 waves (RMS) at 3.5 μm. In parallel with the initiation of this technology program, a cryogenic test facility which can accommodate a SIRTF-sized telescope is under construction at JPL. This facility will be used to test the technology testbed telescope and, perhaps, the actual SIRTF telescope as well.

3.4. Spacecraft Subsystems

SIRTF places no unusual performance requirements on the spacecraft subsystems, although providing the required performance under the combined constraints of minimizing mass and cost and maintaining a high degree of thermal isolation between the spacecraft and the telescope/dewar assembly will require careful design and engineering. Current thinking views the spacecraft not as a stand-alone assembly but as a collection of sub-system modules which are brought together on a common structure. The estimated mass of the current spacecraft configuration - including power, command/data, pointing/reaction control, and telecommunications subsystems - is $\simeq$200 kg. Additional mass reductions may be achievable by using such approaches as the high density, micro-miniaturized electronics packaging now under development for such programs as the Pluto Fast Flyby.

3.5. Integration and Test

The approach to system integration and test is currently being defined. The general philosophy will be to minimize the amount of system-level testing required by doing as much testing as possible at the subassembly and subsystem level. Early experience with the challenges of cryogenic testing of the SIRTF telescope and instruments would clearly be very valuable. For that reason, the test facility is being designed to accommodate not only the full SIRTF telescope assembly, but also the SIRTF instrument chamber, and current plans call for one or more prototype instrument modules to be tested with the telescope as early as 1996.

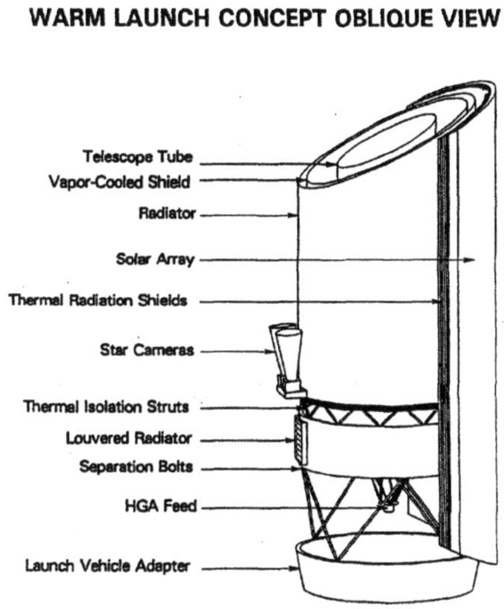

Fig. 4. External view of a warm-launch SIRTF concept. The overall length of the system is about 4.5 meters.

3.6. Summary and Schedule

Although many of the details of the design will remain undetermined until the start of the SIRTF development phase, preliminary analysis has shown that a Delta-launched mission providing the capabilities listed above can be implemented within the available budget. Figure 4 shows one possible configuration for such a spacecraft.

NASA's current plans call for a Phase B start for SIRTF in FY1997. On this schedule, the project development phase would begin in FY1998-99, leading to launch in 2001-2.

4. Instruments and Instrument Technology

SIRTF's three instrument teams are working closely together in the scientific and technical definition of a unified payload. The measurement functionality to be provided by the payload has been determined by the four major scientific themes, and the required functionality and its relationship to these themes is illustrated in Table I. It is envisioned that the SIRTF payload will consist of a number of discrete modules, as described in Table I. These modules will share the SIRTF field of view, with each module being fed by a pickoff mirror or directly, depending on the final design. The module fields of view could project onto the sky as illustrated in Figure 5.

TABLE I
SIRTF Instruments

Imaging Modules

Wavelength μm	Arrays/ Pixels	Field of view (arcmin×arcmin)	Scientific themes
3.5	InSb (256×256)	5×5	Early Universe, Brown Dwarfs
4.5	InSb (256×256)	5×5	Early Universe, Brown Dwarfs
6.5	Si:As - IBC (128×128)[a]	5×5	Early Universe, Brown Dwarfs
8	Si:As - IBC (128×128)[a]	5×5	Early Universe Ultraluminous Galaxies
30	Si:As - IBC (128×128)	5×5	Ultraluminous Galaxies, Early Universe, Planetary Debris Disks
70	Ge:Ga (32×32)	5×5	Ultraluminous Galaxies, Early Universe, Planetary Debris Disks
160	Ge:Ga (1×16)[b]	0.3×5	Ultraluminous Galaxies, Early Universe, Planetary Debris Disks

Spectroscopic Modules

Wavelength μm	Arrays/ Pixels	Resolution[c] $\lambda/\delta\lambda$	Scientific themes
4–5.3	InSb (256×256)	100	Early Universe, Brown Dwarfs
5–15	Si:As (IBC) (128×128)[d]	100	Brown Dwarfs, Planetary Debris Disks, Ultraluminous Galaxies, Early Universe
15–40	Si:Sb (IBC) (128×128)	100	Planetary Debris Disks, Ultraluminous Galaxies, Early Universe
12–24	Si:As (IBC) (128×128)[c]	600	Planetary Debris Disks, Ultraluminous Galaxies
20–40	Si:Sb (IBC) (128×128)	600	Planetary Debris Disks, Ultraluminous Galaxies
55–100[e]	Ge:Ga (1×32)	20	Planetary Debris Disks

[a] 256×256 format under consideration.
[b] Stressed detectors. 2×16 format under consideration.
[c] R=100 modules work in long slit mode for spectral imaging – typical slit lengths several arcmin. R=600 modules work in echelle mode.
[d] 256×256 format under consideration.
[e] Not a separate module, but an operating mode for the 32×32 Ge:Ga imaging array.

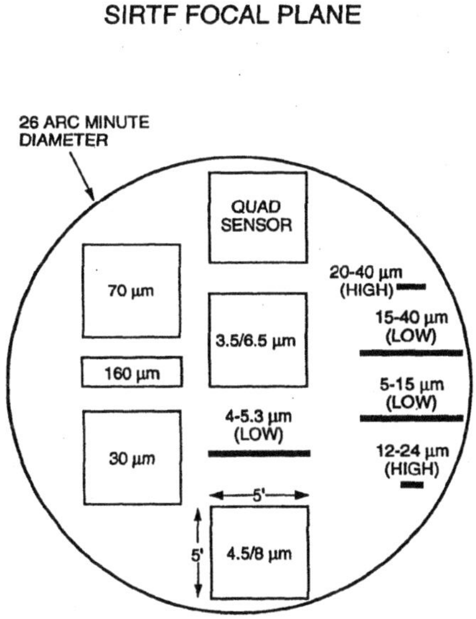

Fig. 5. The projection of the SIRTF instrument apertures onto the sky. The spectrograph apertures include both long slits (LOW) for low resolution imaging spectroscopy and shorter slits (HIGH) for high resolution echelle mode spectroscopy. The two short wavelength fields of view in the center are each shared by two arrays [one InSb, one Si:As] which view the same field by using a dichroic. The quad sensor references the telescope focal plane to the line of sight of the external star tracker by observations of visible stars.

As currently envisioned, the payload described in Table I would contain at most two cryogenic mechanisms - a shutter in the short wavelength camera (required for measurements of dark current), and a dither mirror in the long wavelength camera required for modulation of the signal at $\simeq$0.1 Hz. No mechanisms of the filter wheel or grating drive type are included in the current concept. Thus the imaging modules are restricted to the specific spectral bands indicated in the table, and the spectrometers have no moving parts. Table I also shows the array type(s) and format(s) to be used in each detector module. Note that for a particular spectral band the same material and format will be used for both imaging and spectroscopic applications.

The advantages of detector arrays for imaging are obvious and well-known, but the large format arrays are essential for the no moving parts spectrograph concepts as well. The low resolution spectrographs will use order-sorting filters along the slit to cover more than one spectral order with a single detector array, while the high resolution spectrographs will cover a wide spectral band by operating in echelle format, with a number of spectral orders detected simultaneously.

The technical state of the detector development for SIRTF is summarized in Table II. Two recent developments of particular note are: (1) The demon-

TABLE II
Focal Plane Array Summary

Detector Material	Wavelength Range	Format	Pixel Pitch	Detective Quantum Efficiency	Dark Current	Read Noise
	(Full Width	(Pixels)	(μm)	λ_{peak}	(e^-/sec)	(e^-)
InSb	3–5.5	256×256	30	0.9	< 1	10–15[a]
Si:As	6–28	128×128	75	0.25	10[b]	< 100
Si:Sb	18–38	128×128	75	0.25	260[c]	< 100
Ge:Ga	45–115	32×32	750	0.25	< 150	50
Ge:Ga stressed	115–200	1×16	6000	0.07	100	65

[a] With best multiple sampling techniques.
[b] Improvements in design and processes should yield 5 e^-/sec.
[c] Improvements in design and processes should yield 40 e^-/sec.

stration of a 1×32 cryogenic readout device which operates at <5 K with a power dissipation of <1 μW per channel. A 1×32 linear detector array read out by such a device will be the basic module from which the 32×32 Ge:Ga imaging array will be assembled. The availability of a low power, low temperature readout will reduce the risk and complexity of the assembly of this array while bringing its power dissipation down to a level consistent with the capabilities of the moderate-class SIRTF. The second noteworthy development has been the demonstration of Si:Sb IBC devices as viable detector arrays for the 15-40 μm region. This development makes efficient imaging and spectroscopy possible in a largely unexplored spectral band.

With these developments in hand, the fundamental technologies for SIRTF's detectors have been demonstrated, and the basic performance requirements for the detectors have been met or exceeded across the spectral band. The emphasis of the detector work is now shifting towards engineering-oriented issues such as fabrication yield, operability, strategies for dealing with ionizing radiation, and flight packaging and cryogenic cabling.

5. Science Program and Science Operations

The scientific rationale and science objectives for SIRTF have been described in several recent publications, to which the reader is referred for a detailed discussion (Werner, 1993, 1994; Cruikshank, Werner, and Backman, 1994). The scientific significance of the four themes listed above - and the scope of SIRTF's potential contributions - is summarized very briefly below.

5.1. Protoplanetary and Planetary Debris Disks

The discovery of disks of particulate matter orbiting solar type stars was one of IRAS' major discoveries. With the notable exception of the disk around Beta Pictoris, these systems are not seen at visible wavelengths. They were readily detectable by IRAS because they absorb and reradiate a small fraction of the light from the central star but are in general too faint to be seen with ground-based or airborne infrared telescopes. Modelling of the IRAS data suggests that these disks contain particles ranging in size from $\simeq 1\,\mathrm{mm}$ up to bodies as large as asteroids and comets, with some arguments suggestive of still larger, planet-sized objects. Thus further exploration of this phenomenon will help us to understand the frequency and properties of planetary systems around nearby stars.

SIRTF can contribute significantly to this exploration by detecting debris disks around a statistically meaningful sample of stars, so that the frequency of occurrence and properties of the disks can be correlated with the stellar properties. Detailed imaging of the nearby disks can constrain models for the grain properties and distribution while searching for the central voids inferred from the IRAS data. In addition, SIRTF can compare low resolution spectra of the circumstellar material with spectra of emission from comets in our own solar system in order to compare the physical conditions in the primitive solar system to those in the debris disks.

5.2. Brown Dwarfs and Superplanets

Brown dwarfs are as-yet hypothetical objects with masses between 0.001 and $0.08\,M_\odot$ which are too low in mass to sustain nuclear burning but may be visible in the infrared as they radiate the gravitational energy released in their formation. They are of particular interest because they may account for some of the "missing mass" which forms a dynamically significant but as yet unseen halo for our Galaxy - and other galaxies as well. As brown dwarfs age, they become cooler and fainter in the infrared. However, in a single 600 second integration at $4.5\,\mu$m, SIRTF can detect $0.03\,M_\odot$ objects with ages of 10 billion years - as is appropriate for the halo of our Galaxy - at distances up to 40 pc from the sun. Brown dwarfs are one of the potential targets of the SIRTF Key Project surveys described below. For example, the local density of a dynamically-significant halo would be about $0.01\,M_\odot\,\mathrm{pc}^{-3}$. If this mass is in the form of $0.03\,M_\odot$ objects, approximately 1000 brown dwarfs should be present in the data base created by a SIRTF survey which covers $\simeq$1% of the sky to the sensitivity limit indicated above.

5.3. Ultraluminous Galaxies and Active Galactic Nuclei

IRAS provided our first complete look at the extragalactic sky in the infrared. In addition to showing that $\simeq$50% of the luminosity of a typical spiral galaxy is radiated at far infrared wavelengths, IRAS identified several new classes of infrared-luminous galaxies. One such class which was found in the main IRAS catalogs is the ultra-luminous galaxies - those having $L > 10^{12}\,L_\odot$. These have optical images suggestive of violent interactions and mergers and luminosities well into the luminosity range of quasars. More recently, several even more luminous galaxies - one of which, FSC10214, which has $z > 2$ and $L > 10^{14}\,L_\odot$, is among the most luminous objects known in the Universe - have been found through followup studies of IRAS' Faint Source Catalog and Data Base. Like the ULG's, the FSC10214 objects radiate 90 to >99% of their total luminosity at infrared wavelengths.

SIRTF can detect ultra-luminous galaxies to $z > 2$ and study their evolution in space-time. If ultra-luminous galaxies are triggered by galaxy interactions, their number density should have been proportionately greater at earlier epochs, when the higher co-moving density would have led to more frequent galaxy-galaxy interactions. The same observations would detect FSC10214 class objects to $z > 5$ and determine whether they are a distinct population - perhaps powered by a different physical mechanism - than the lower luminosity ULG's. Finally, SIRTF's spectrographs can measure emission lines from different ionic species to determine the physical conditions in the optically-obscured interiors of these objects, thereby distinguishing between stellar and non-thermal processes as the ultimate source of their luminosity.

5.4. The Early Universe

SIRTF's deep imaging at wavelengths between 3 and 8 μm, which would be capable of detecting L* galaxies to $z > 3$, can probe the early universe. These measurements permit a photometric redshift determination, because the position of a galaxy in an infrared color-color diagram depends on the redshifted position of the broad flux maximum which appears at 1.6 μm in the galaxy rest frame. This feature is due to the minimum in the H-opacity at this wavelength, and it should be a persistent feature in the spectral energy distribution of any stellar population dominated by K and M giants in the infrared. Comparison of SIRTF's census of galaxy number and properties as a function of redshift out to $z > 3$ with the results of model calculations will test our understanding both of the evolution of galaxies and of the geometry of space-time.

The investigations of the most luminous infrared galaxies discussed earlier will provide additional insights into the early universe. It has been suggested that these objects may be protogalaxies - undergoing an initial cataclysmic burst of star formation - because their luminosity is far too high to be sustained by nuclear burning in a normal stellar population for even a small fraction of the Hubble time. If this conjecture is true, the numbers of such objects ought to increase markedly with look-back time, and SIRTF's ability to detect them to $z > 5$ and beyond ought to reveal many new examples.

5.5. Other Scientific Themes

Although only the four major science programs described above have been permitted to drive the cost and complexity of SIRTF, the mission defined in this way will have powerful capabilities for the study of many other scientific question. Examples include:

1. The early stages of star formation - SIRTF can identify cold condensations signalling the earliest onset of gravitational fragmentation in molecular clouds;
2. The origin and dissemination of the chemical elements - SIRTF can monitor the return of freshly-synthesized material to the interstellar medium by spectroscopic studies of evolved stars, planetary nebulae, novae, and supernovae;
3. The outer solar system - SIRTF can determine the sizes of Kuiper belt planetesimals discovered optically and search for objects of even lower albedo which might be overlooked in the visible surveys.

5.6. Science Operations

The redefined SIRTF described here will continue to be an observatory accessible to the entire scientific community. However, the reduced lifetime and operations phase funding have sharpened the requirement that SIRTF science operations be carried out both with high efficiency and in a manner which creates a coherent scientific legacy which will support a productive archival research program. At the same time, the nature of the scientific themes ennunciated for SIRTF above calls for a scientific program which emphasizes large scale projects - targeted and unbiased spectroscopic and imaging surveys - early in the mission to identify promising objects for detailed follow-on study.

These considerations have led to the definition of a science program for SIRTF which is centered on Key Projects - such as the large scale surveys alluded to above. These projects and the investigators to implement them would be selected well in advance of through a peer review process open to the entire scientific community. The selected Key Project investigators would work closely with the SIRTF Science Working Group, instrument teams, and science operations staff before launch. In addition to the Key Projects, the SIRTF science program will include a Guest Observer program which provides another means of community participation. Additional components of the science program are the Guaranteed Time Observations, carried out by the Science Working Group and the instrument teams, and a flexible means of responding to Targets of Opportunity.

5.7. Serendipity

The results of IRAS have revalidated the astronomical truism that major advances in capability lead to unexpected and exciting discoveries. Thus the most important scientific results from SIRTF are likely to result from serendipitous discoveries which cannot be foreseen in advance but which the science operations system must anticipate. Such serendipitous discoveries may be found in the large scale surveys which will be carried out under the Key Project program described above. An important part of this discovery process will be the followup observations, some of which can only be carried out from SIRTF. One of the means which has been proposed to facilitate this followup would be to declare the data from the Key Projects non-proprietary, so that it is released to the scientific community as soon as it is available in scientifically useful form. Means such as this will be required to maximize the followup opportunities during SIRTF's short lifetime.

6. Conclusion

Technical innovation in the facility design and the continually improving performance of infrared detector arrays have kept SIRTF scientifically viable even as the available resources have fallen well into the moderate mission class. Recently, the Committee on Astronomy and Astrophysics, a group charged by the US National Academy of Sciences with tracking the status of the Bahcall Committee recommendations, has stated that the moderate mission SIRTF described here still merits the high-priority rating it received from the Bahcall Committee. We are eager to move this mission forward - primarily for the tremendous scientific return which it will yield - but also as a positive example of NASA and the scientific community's response to a new, resource-constrained approach to space science.

Acknowledgements

This work was carried out at the Jet Propulsion Laboratory, California Institute of Technology, under contract with the National Aeronautics and Space Administration

References

Cruikshank, D.P., Werner, M.W., Backman, D.E. 1994: *Astrophys. Sp. Sci.*,**212**, 407
Hawarden, T.G., Cummings, R.O., Telesco, C.M., Thronson, Jr., H.A. 1992: *Sp. Sci. Reviews*, **61**, 113
Werner, M.W. 1993: *Astr. Soc. Pacific Conf. Series*, **43**, 249
Werner, M.W. 1994: *Infrared Phys. Technol.*, **35**, 539

THE EDISON INFRARED SPACE OBSERVATORY

H. A. THRONSON
Wyoming Infrared Observatory, University of Wyoming, USA
T. G. HAWARDEN
Joint Astronomy Centre, USA/Royal Observatory, Edinburgh, UK
A. J. PENNY
Rutherford Appleton Laboratory, UK
L. VIGROUX
Service d'Astrophysique, Saclay, France
and
G. SHOLOMITSKII
Space Research Institute, Moscow, Russia

Abstract. For five years, the *Edison* program has had the goal of developing new designs for infrared space observatories which will "break the cost curve" by permitting more capable missions at lower cost. Most notably, this has produced a series of models for purely radiative and radiative/mechanical ("hybrid") cooling which do not use cryogens and optical designs which are not constrained by the coolant tanks. Purely radiatively-cooled models achieve equilibrium temperatures as low as about 20 K at a distance of 1 AU from the sun. More advanced *Edison* designs include mechanical cooling systems attached to the telescope assembly which lower the optical system temperature to $\simeq$ 5 K or less. Via these designs, near-cryogenic temperatures appear achievable without the limitations of cryogenic cooling. One *Edison* model has been proposed to the European Space Agency as the next generation infrared space observatory and is presently under consideration as a candidate ESA "Cornerstone" mission. The basic design is also the starting point for elements of future infrared space interferometers.

Key words: infrared – *Edison* – space observatories – radiative cooling

1. Background: Why Infrared Astronomy from Space?

Infrared ($\lambda \approx 1 - 200\,\mu$m) observations provide fundamental information about the structure, composition, state, evolution, and energetics of astronomical sources as close as planets within our Solar System and as distant as putative primeval galaxies. This is simply a result of the fact that a major fraction of all diagnostic spectral features, as well as much of the total luminosity from many sources, are emitted at infrared wavelengths. In addition, due to universal expansion, the bulk of the radiation emitted by very distant objects at shorter wavelengths in their rest frame are observed beyond about $1\,\mu$m from the Earth. It is for these reasons that scientific advisory committees in Europe, the US, and elsewhere have identified this wavelength range as of paramount importance in the future study of the cosmos.

Unfortunately, most of the infrared regime is inaccessible from the surface of the Earth due to atmospheric obscuration. Furthermore, in those important windows accessible from within the Earth's atmosphere, faint signals

Space Science Reviews **74**: 139–144, 1995.

from space are overwhelmed by emission from both telescope optics and the atmosphere itself. Therefore, although useful infrared observations are possible both from high mountain tops and from airplanes and balloons in the stratosphere, celestial-background-limited sensitivity is possible only from space.

2. A Modern Infrared Space Observatory

Over the past few decades, a series of missions have been proposed to undertake increasingly sophisticated infrared observations from space. Most familiar to civilians not working in this field is a basic cryogenically-cooled system of the type adopted for the successful *Infrared Astronomical Satellite* (IRAS) mission about a decade ago, as well as ESA's *Infrared Space Observatory* (ISO) due to be launched in late 1995. In this design, both the instruments and the telescope are surrounded by a toroidal tank of liquid helium, which effectively lowers the optical system temperature to a few degrees kelvin. These designs are limited, however, both in the lifetime and the telescope aperture permitted by the cryogen tanks. In the US, NASA's *Space Infrared Telescope Facility* (SIRTF) program probably pushed this type of design to its limit within the confines of realistic budgets, producing a 0.85 m telescope and associated sophisticated instrument system (e.g., Garcia 1993).

However, the overall performance of a telescope increases rapidly with aperture, so that there are very good reasons to try to break the limits imposed by one type of technology and to significantly increase the telescope's collecting area. For example, for point sources, the time that it takes to reach a given flux level declines as (telescope diameter)$^{-4}$. Therefore, increasing a telescope's diameter by a factor of 3 (equivalent to the increase of the baseline *Edison* design over that of ISO) means a decline in integration time of almost 2 orders of magnitude. If the same technology also allows a much longer operational lifetime, then the effective performance of an observatory will be further multiplied.

For those reasons a number of workers over the past decade have produced increasingly detailed alternative designs which take advantage of the low temperature of space to cool large portions of the optical system. The ISO Pre-Phase A study in 1980 appears to have been the first civilian proposal for an extensively radiatively-cooled observatory, although subsequent designs for this spacecraft opted instead for an IRAS-like cryogenic cooling system. About a decade later, a team of scientists at the Royal Observatory Edinburgh led by T. G. Hawarden proposed to ESA the innovative *Passively-cooled Orbiting Infrared Observatory Telescope* (POIROT), which was the first major civilian proposal to do without cryogens entirely, replacing them with a mixture of radiative and mechanical cooling. This was a watershed in IR space astronomy, as henceforth IR space missions need not be so severely limited in both aperture and lifetime for a given cost (see also Lin 1991, Rapp 1992). The POIROT proposal to ESA, although not funded by the European agency, led directly to the more ambitious international *Edison* program. This was the first proposed IR space observatory with the stated goal of using new technologies to permit both larger aperture and longer lifetime, rather than very low temperatures, as had been the case with previous proposed IR space observatories. Historical developments of non-cryogenic cooling in space observatories are described in more detail in the papers by Hawarden et al. (1992 and these proceedings) and Thronson et al. (1992).

3. *Edison*: The Baseline Design

Edison's dominant mission requirement is to maintain as large a telescope aperture as permitted by available launch vehicle and budgets, while using appropriate (and innovative) technologies to cool parts of the facility to temperatures necessary for extremely sensitive observations in the infrared. As Thronson et al. (1995) showed, these temperatures do not always need to be as low as those achieved via cryogenic cooling. An external view of the current baseline *Edison* design is presented in Figure 1 which has a 1.7 m f/20 Cassegrain telescope surrounded by a set of nested radiation shields. The proposed orbit for this design is a "halo" around the L2 point, which allows a pair of sunward radiation shields which place the telescope in deep shade. In addition, segments of the concentric radiation shields can be cut away on the anti-sunward side to expose large areas of the telescope structure directly to the cold of space. This design adopts the same service module as that under development for ESA's Solar and Heliospheric Observatory (SOHO) program as it meets the majority of mission goals and recovers some of the cost of development of this other project.

The baseline optical design was produced by Eastman Kodak and is based upon a 1.7 m f/1.2 fused silica primary as part of a f/20 Cassegrain design. The tertiary is proposed to be a four-faced pyramid, feeding four instrument bays. Following ESA guidelines, these instruments are to be contributed by national groups, which allows an opportunity for international collaboration. An on-axis guide camera is located behind a hole in the vertex of the pyramid. We estimate that the field of view of each instrument will be about $5'$.

The very lightweight primary has eggcrate cut-outs produced by Kodak's abrasive water jet system. The mass of the 1.7 m telescope system is about 420 kg. An 0.5 m test mirror of this design was ion figured about a decade ago by Kodak to about 0.1 wave (optical) and cooled to 8 K with minimal loss of figure. The baseline mirror coating is vacuum-deposited gold, which has a reported emissivity below $\epsilon = 0.01$ at low temperatures at wavelengths

Fig. 1. An external view of the *Edison* spacecraft, which contains a 1.7 m telescope within an overall spacecraft not much larger than ESA's *Infrared Space Observatory*. As the observatory is proposed to occupy a "halo" orbit around the L2 point, a pair of solar panels/radiation shields occupy one side and large areas are cut away from the anti-sunward portion. Via radiative cooling alone, this design achieves an equilibrium temperature of about 20 K, although mechanical cooling systems are proposed to lower the optical system temperature to around 5 K.

beyond about 10 μm. Therefore, it may be possible to produce an optical system with total emissivities at far-infrared wavelengths of a few percent or less.

In this design, *Edison* avoids the use of cryogens in any capacity. A variety of theoretical calculations indicate that optical system temperatures in the range of about 20 K are possible via radiative cooling alone for designs similar to that shown in the figure. After a cooldown from room temperature to about 30 K, which takes about 3 months in the current design, a pair of Stirling cycle mechanical coolers are turned on, which takes the telescope and instrument system down to about 20 K in a few days. These Rutherford Appleton Laboratory coolers may also be used to stabilize the optical system temperature against variations in external heating due to changes in observatory orientation or degradation of spacecraft surface properties.

However, 20 K is not sufficiently cold for sensitive operation of current detectors beyond about 5 μm. Our baseline design therefore incorporates a suite of Stirling and mixed Joule-Thomson/Stirling mechanical coolers to achieve even lower temperatures. These types of systems are being adopted for a variety of space missions and have extensive laboratory experience. The types of coolers appropriate to *Edison* will probably be space qualified within a few years.

4. *Edison*: Possible Future Design Developments

If approved by ESA for further study, the *Edison* program will continue its central activity of pursuing innovative technology leading to the most capable space observatory within reasonable budgets. With many years — even decades — before a plausible launch of the mission, it is important to continue to develop new designs and/or new technology which will allow improved performance.

The basic concepts behind radiative cooling for large optical structures are outlined by Hawarden et al. elsewhere in this volume. These authors argue that temperatures in the range of about 20 K are likely to be about the lowest possible via radiative cooling alone at about 1 AU from the sun, given realistic assumptions about input power from instrument systems. Although this temperature range is substantially below that considered possible just a few years ago, this will still produce a substantial thermal background from the optical system at wavelengths longward of about 50 μm. This will reduce the overall system sensitivity substantially (e.g., Thronson et al. 1995). As a consequence, the *Edison* team has developed a series of alternative designs which will allow lower optical system thermal background or form the basis for special-purpose IR space missions, such as the US Small Explorers (SmEx) and Mid-sized Explorers (MidEx)

One advanced design, the so-called Very Cold Telescope (VCT), uses mechanical refrigeration systems to cool not only the detectors, but also the entire optical system to temperatures approaching that of liquid helium (Hawarden et al., these proceedings). Current estimates are for a telescope which will equilibrate at about 5 K. This means that cryogenic temperatures may be achieved without the limitations — lifetime and constrained aperture — of these coolants. In addition, as some mechanical cooling systems are microphonic, cooling the entire telescope, rather than only the detectors, may allow a significant reduction in potential mechanical and electrical noise.

In addition to lowering the optical system temperature, the *Edison* team is exploring alternative optical designs which may significantly lower overall system emissivity. Work on the French PRONAOS sub-millimeter balloon-borne telescope suggests that mirror emissivity at infrared wavelengths can be below $\epsilon \approx 0.01$ at low temperatures. Off-axis/non-axisymmetric optical designs are well-suited to take advantage of very low mirror emissivities. As this is being written, an *Edison* program has begin to produce a design for an infrared space observatory with an overall system emissivity of $\epsilon \leq 0.01$.

Cryogenic space missions (e.g., Garcia 1993) can be designed somewhat similarly to groundbased optical observatories, where there is no telescope emission at the wavelengths of operation. However, radiatively-cooled space observatories instead adopt optical designs more like those of groundbased IR telescopes, attempting to reduce potential sources of emission throughout the system. An off-axis/non-axisymmetric design with very low-emissivity optics may permit a large aperture in a compact space and may not require a very low temperature for sensitive operation. Such a design has been proposed as one alternative for the ESA FIRST "Cornerstone" mission (e.g., Thronson et al. 1995).

Some of the results of the *Edison* program are being incorporated in a program to design a *ground-based* very cold telescope which will reduce the optical system emission to well below that of the atmosphere. For example, there appear to be a few very transparent bands in the 8 – 12 μm range where a high transparency and very low telluric background might be achieved. A very cold ($T \leq 70\,\mathrm{K}$) telescope at one of the best mountain sites or in Antarctica might achieve near-space sensitivity.

Finally, these basic *Edison* designs may be used as the starting point for elements of a future infrared/sub-millimeter array in orbit (e.g., Léger et al., these proceedings). Such a telescope appears to be required for high spatial resolution astrophysical programs, including detection of Earth-like planets around stars.

Acknowledgements

Design work on the next generation infrared space observatory at the University of Wyoming has been supported by the University Research Office, the NASA Space Grant program, and NSF grant AST 91-17096. This work has also been supported in Europe by Rutherford Appleton Laboratory, the Joint Astronomy Centre, and the Royal Observatory, Edinburgh. The authors deeply appreciate the support of their colleagues during the years when the ideas summarized here were heretical.

References

Garcia, M (ed.): October, 1993, *Space Infrared Telescope Facility – Mission Concept [JPL D-11183]*, Jet Propulsion Laboratory: Pasadena

Hawarden, T. G., Cummings, R. O., Telesco, C. M., Thronson, H. A.: 1992, *Space Sci Revs* **61**, 113

Lin, E. I.: July, 1991, *A 10-meter 20-Kelvin Infrared Space Telescope: The Passive Cooling Approach [Proposal]*, Jet Propulsion Laboratory: Pasadena

Rapp, D.: March, 1992, *Potential for Active Structures Technology to Enable Lightweight Passively Cooled IR Telescopes [JPL D-9449]*, Jet Propulsion Laboratory: Pasadena

Thronson, H. A. et al: 1992, *Space Sci Revs* **61**, 145

Thronson, H. A., Rapp, D., Bailey, B., Hawarden, T. G.: 1995, *PASP, in press* ,

THE BOOMERANG EXPERIMENT

A. LANGE
California Institute of Technology, Pasadena, USA

P. DE BERNARDIS, M. DE PETRIS, S. MASI and F. MELCHIORRI
Dipartimento di Fisica, Universita' La Sapienza, Roma, Italy

E. AQUILINI, L. MARTINIS and F. SCARAMUZZI
ENEA, Frascati, Italy

B. MELCHIORRI
IFA-CNR, Roma, Italy

A. BOSCALERI
IROE-CNR, Firenze, Italy

G. ROMEO
Istituto Nazionale di Geofisica, Roma, Italy

J. BOCK, Z. CHEN, M. DEVLIN, M. GERVASI, V. HRISTOV, P. MAUSKOPF, D. OSGOOD and P. RICHARDS
Physics Department, U.C.B., Berkeley, USA

and

P. ADE and M. GRIFFIN
Queen Mary and Westfield College, London, UK

Abstract.
The BOOMERANG (Balloon Observations Of Millimetric Extragalactic RAdiation aNd Geophysics) experiment is an international effort to measure the Cosmic Microwave Background (CMB) anisotropy on angular scales of $20'$ to $4°$, with unprecedented sensitivity, sky and spectral coverage. The telescope will be flown from Antarctica by NASA-NSBF with a long duration stratospheric balloon (7-14 days), and is presently scheduled for flight in 1995-1996. The experiment is designed to produce an image of the Cosmic Microwave Background with high sensitivity and large sky coverage. These data will tightly constrain the baryon density, the reionization history, and the formation of large-scale structure in the universe. BOOMERANG will test technologies and return science data that are essential to the design of a future space-borne mission to map CMB anisotropy.

1. The Rationale

Two years of data from the COBE-DMR satellite (Bennet et al., 1994) provide a $10\,\sigma$ detection of the *rms* CMB anisotropy: $30\,\mu$K when observed through a $10°$ FWHM beam. Many cosmological scenarios are consistent with the COBE measurement. A precise measurement of the power spectrum of CMB anisotropies at smaller angular scales, beyond the resolution of COBE, can distinguish between these scenarios. Angular scales of 1 degree correspond to the largest structures that have been mapped in the current epoch, and thus allow a direct comparison of initial and final conditions. Anisotropies at angular scales $< 1°$ are produced by the Doppler effect, and correspondingly, a "Doppler peak" is expected in the power spectrum. The

Space Science Reviews **74**: 145–150, 1995.

amplitude and the position of the Doppler peak probe the baryon density and the reionization history of the Universe.

There is a statistical variance in the power spectra of CMB anisotropy measured by observers located in different regions of the universe. This effect is called Cosmic Variance. Since we can, at best, measure only one manifestation of the statistical ensemble of CMB skies, our ability to constrain theory is ultimately limited by this Cosmic Variance. The COBE-DMR results are presently limited by Cosmic Variance. The Cosmic Variance is smaller at smaller angular scales, because there are a larger number of regions on the sky that can be considered independent samples of the parent distribution. All of the experiments to date (see the review by White et al., 1994) have sampled very limited regions of the sky. The precision with which the amplitude of the power spectrum can be determined by these experiments is thus limited not by Cosmic Variance but by the small size of the sample (Sampling Variance). The precision can be greatly improved simply by increasing the number of pixels that are observed. We have performed a simple simulation to quantify these effects (see de Bernardis et al., 1994, for details). We have used 1000 realizations of CMB anisotropy maps generated in the framework of the CDM model (n=1, COBE-DMR normalization, $\Omega_B = 0.05$). We have computed, for each realization, the ΔT_{rms} measured sampling the sky, and have plotted the results in histograms. We have considered three different experiments: (a) a short balloon flight, similar to present anisotropy experiments, measuring ΔT for 33 sky patches with a beam 40 arcmin FWHM; (b) a long balloon flight, measuring a CMB anisotropy map $10° \times 10°$ wide with 20 arcmin FWHM ($\simeq$ 900 ΔTs, as planned for BOOMERANG); (c) a dedicated CMB anisotropy satellite, measuring a larger map with a total of 3600 ΔT (this is a rough estimate of what is possible to do without significant foreground contamination). In the three cases it has been assumed that both detector noise and astrophysical confusion noise are negligible: the spread of the histograms is due only to the combined effect of Cosmic and Sampling Variance. The FWHM of the histograms is 20 μK, 11 μK and 8 μK in cases (a),(b),(c) respectively. It is evident from these results that experiment (b) can, in principle, make a significant improvement on experiment (a) and can, in fact, come close to what is ultimately possible with an orbital mission. We emphasize that this is the case only if the experiments have adequate sensitivity and frequency coverage to make detector noise and astrophysical confusion noise negligible in the available amount of observing time and sky coverage. We consider Case b) as the design goal for the BOOMERANG experiment. BOOMERANG is designed to minimize galactic foreground confusion by mapping the region of the sky that is least contaminated with galactic foreground emission with unprecedented frequency coverage, from 40 GHz to 450 GHz. BOOMERANG will map a $10° \times 10°$ region of the sky with better than 0.5 degree resolution and a final

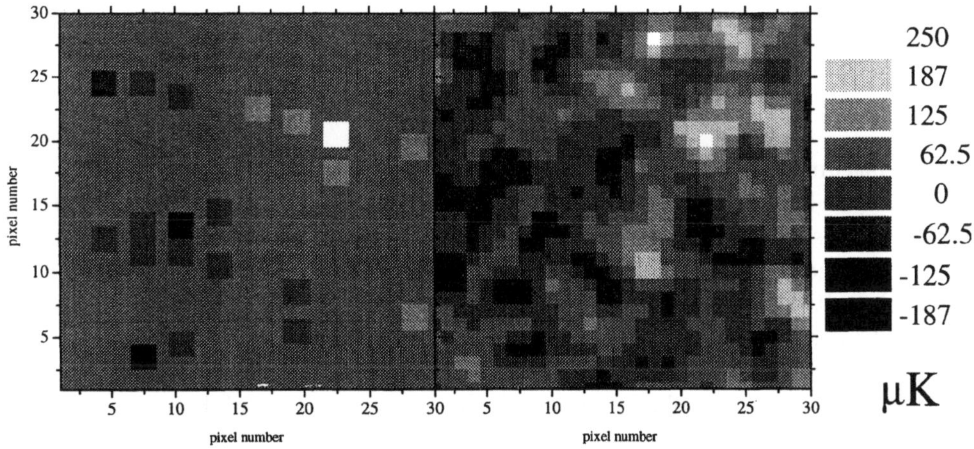

Fig. 1. The same CMB anisotropy simulation, as measured using a present day CMB anisotropy experiment (left) and the BOOMERANG experiment (right). The size of the map is $10° \times 10°$; the pixel size is 40 arcmin in the case on the left, 20 arcmin for BOOMERANG.

uncertainty/pixel in CMB temperature of about $3\,\mu$K. A visual comparison of the same simulated CMB anisotropy map, observed with present day experiments and with BOOMERANG, is shown in Figure 1. The achievable gain in information is evident: while present day experiments can only derive informations on the *rms* ΔT in the map, the BOOMERANG experiment will yield detailed information about the topology of the map.

2. Why from Antarctica?

There are two unique features of Antarctic Ballooning which make it very attractive for CMB anisotropy experiments. The first one is the flight duration. Extensive tests made by NASA-NSBF have shown that stratospheric balloons launched from latitudes $\simeq -80°$ travel along the meridian, coming back to the launch site in $7 - 20$ days. This has to be compared to $\simeq 1$ day flights available from temperate latitudes. The second feature is that the launch campaign is performed during the Antarctic summer (December and January) and a region with very low Galactic dust contrast is available in the anti-sun direction. The region, as it appears in the IRAS Sky Survey Atlas, is at R.A. $\simeq$ 4.5 hours, dec. $\simeq -45°$, and is roughly 300 square degrees in area. The dust brightness contrast is < 0.2 MJy/sr at $100\,\mu$m, the lowest of any region this size. If we use a thermal dust spectrum with $n \simeq 1.5$ and $T_d \simeq 21$ K we get a scaling of $\simeq 5\,\mu$K/(MJy/sr) in the 4-7 cm^{-1} band.

This level of contrast in the dust emission corresponds to CMB temperature anisotropy of less than $3\,\mu$K.

3. The Instrument

The instrument design evolved from the MAX (Fischer et al., 1992), the ARGO (de Bernardis et al., 1990, 1993) and SUZIE (Wilbanks et al., 1994) experiments. The main components of the experiment are 1) an off-axis 1.2 m telescope 2) a 4-band $\times$ 8 pixel bolometric receiver 3) an alt-azimuth gondola with reaction wheel pointing capability. The instrument has been specially designed to overcome problems typical of Antarctic ballooning: the long flight duration forces us to use special cryogenic systems; the enhanced cosmic rays flux in polar regions forces us to use special bolometers; the presence of the sun allows us to use solar panels for power supply but also forces us to use multiple sun shields for good thermal performance of the system; the balloon is far from the ground equipement, so special data collection/telemetry systems have to be used, and interactivity with the system is reduced. The concept for the experiment is shown in Figure 2.

Here we can only list some technical highlights of the system:

Bolometers have been developed in Berkeley with "spider web" absorbers, that have very high absorption at mm wavelengths and very low geometrical cross section to cosmic rays (see Bock et al., 1994). The bolometers are AC biased (Wilbanks et al., 1994). The very good low-frequency performance allow electronic differencing between pixels and/or observation in total power mode. This strategy, which is essential for an orbital mission, eliminates the need for a mechanical chopper, and with it problems of microphonics and reliability during the long flight. The bolometers are assembled in four-band photometers, with a total of 8 pixels observing simultaneously the bands 2.5-4 cm^{-1}, 4-7 cm^{-1}, 7-10 cm^{-1}, 13-15 cm^{-1}. The filter bands are defined by resonant metallic meshes developed in London. The first band has been optimized to monitor Galactic free-free, the second and third ones are CMB anisotropy channels, the fourth one is a monitor for thermal emission from Galactic dust. Two pixels in the center of the focal plane observe in a fifth band at 1.2-2.8 cm^{-1}.

The *optical system* features a 1.3 m, ambient temperature, off-axis parabolic primary, and a re-imaging optics box at 1.6 K, inside the cryostat, composed by an ellipsoidal secondary and a paraboloidal tertiary. Winston concentrators define the acceptance of the 8 photometers. All the 10 photometer beams overlap within 0.1% on the primary mirror, reducing the sensitivity of the system to thermal and emissivity gradients on the primary. The location of the 10 pixels in the sky is sketched in Figure 2: the total power signals can be differentiated during the post flight analysis in any combination, thus obtaining a number of different filter functions with a single experiment.

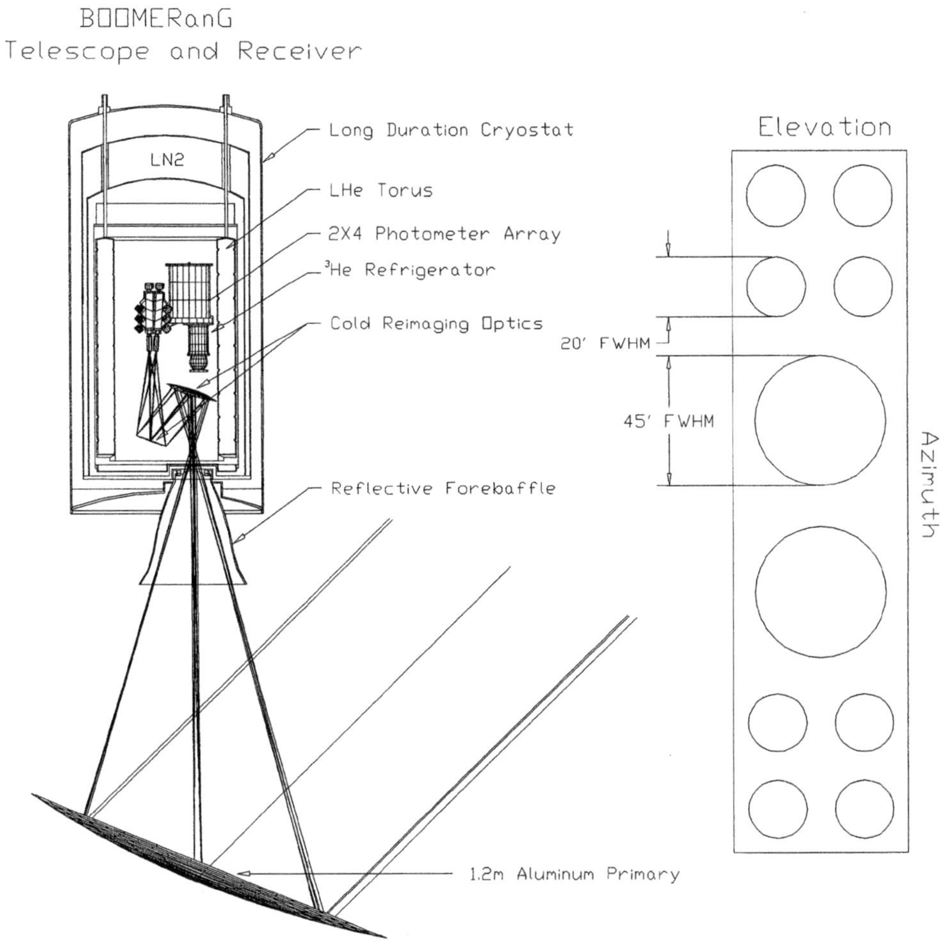

Fig. 2. The concept of the BOOMERANG detection system. In the box is a sketch of the eight photometric pixels projected in the sky. Each pixel contains four spectral bands.

The bolometers are cooled at 300 mK using a single shot ^{3}He fridge in a long duration ^{4}He *cryostat* developed in Frascati and Roma. This consists of a 70 liter liquid N_2 tank and a 40 liter liquid ^{4}He tank. Thermal insulation is achieved by supporting all the tanks with Kevlar cords. Aluminized mylar superinsulation is provided between the 300 K and the 77 K stages, while a vapor cooled 35 K shield is mounted between the 77 K and the 2 K stages. The hold time of the system is 20 days. The large size of the tanks arises mainly from both the large volume of the photometers array and from the presence of a large focal plane, forcing us to accept the thermal radiation input from a large optical window. The only modulation of the optical signal is achieved by scanning the full gondola in azimuth at approximately one scan/minute. The *pointing system*, developed in Florence and Rome, is capable of a pointing accuracy of one arcmin, and of a scan speed of 0.5 deg/sec with a power supply of $\simeq 60$ W. We use sun sensors (a rough one with 360 deg range and a fine one with 120 deg range) to provide the error signals for the feedback loop controlling the inertia wheels. No moving parts are present in the sensors, again gaining reliability for long duration operations. The *final sensitivity* of the system can be estimated as follows. We assume 10 days of useful data in a 14 days flight; this gives us about 1 hour of observation for each of the 900 pixels in the map if 4 photometric pixels are used. The electrical NEP of the AC biased spider bolometers is of the order of $2 \times 10^{-17}\,\mathrm{W}/\sqrt{\mathrm{Hz}}$, with an overall efficiency of the photometers of $\simeq 10\%$. This translates, in the 4-7 cm^{-1} band, in a $NET \simeq 150\mu\mathrm{K}/\sqrt{\mathrm{Hz}}$ for the thermodynamic temperature fluctuations of a 3 K blackbody. As a result the final sensitivity of the experiment is $\simeq 3\,\mu$K per pixel. Note that each sky pixel is observed by all the photometric pixels in our array: this gives us redundancy and allows us to make important consistency tests in the data set. We are presently building most of the subsystems of the experiment, which is planned for a flight in 1995/1996.

Acknowledgements

The BOOMERANG experiment is supported in the U.S. by NASA IR branch, by the NSF Polar program and by a seed grant from NSF Center for Particle Astrophysics; by PPARC (grant GR/J 86599) in the UK; by PNRA, ASI and MPI in Italy.

References

Bennet, C.L., et al.: 1994, 'Cosmic Temperature Fluctuations from two years of COBE DMR observations', *COBE preprint* **94-01**,
Bock J., et al.: 1994, *these proceedings* ,
de Bernardis P., et al.: 1990, *ApJLett* **360**, L31
de Bernardis P., et al.: 1993, *A&A* **271**, 683
de Bernardis P., Muciaccia F., Vittorio N.: 1994, *in preparation* ,
Fischer M.L., et al: 1992, *ApJ* **388**, 242
Fischer M.L., et al: 1994, 'Measurement of the millimeter-wave spectrum of interstellar dust emission', *ApJ* , in press
White M., Scott D., Silk J.: 1994, 'Anisotropies in the Cosmic Microwave Background', *ARA&A* ,
Wilbanks T., et al: 1990, *IEEE Trans. Nuclear Science* **37**, 566
Wilbanks T., et al: 1994, *in preparation* ,

THE FAR INFRARED EXPLORER (FIRE)

A.E. LANGE and J.J. BOCK
Dept. of Physics, University of California, Berkeley, CA

and

P. MASON
Jet Propulsion Laboratory, Pasadena, CA

Abstract.
We describe the Far IR Explorer, a MIDEX-class orbital mission designed to survey the entire sky at millimeter and sub-millimeter wavelengths. The primary science goal of FIRE is to map the Cosmic Microwave Background with $20'$ resolution and 1 ppm precision. In addition, FIRE will measure diffuse radio and infrared emission from the Galaxy with unprecedented sensitivity, and will uniformly survey the entire sky to a limiting flux density of < 100 mJy ($3\ \sigma$).

1. Introduction

The Cosmic Microwave Background (CMB) is a relic of the primeval fireball that filled the universe at the time of the Big Bang. Less than a million years after the Big Bang, the universe cooled sufficiently for the plasma of electrons and nuclei to combine to neutral form. This "recombination" decoupled the CMB from matter. The spatial distribution of the CMB thus reflects the distribution of matter and energy on a photosphere close to the edge of the observable universe, before baryonic matter could gravitate into the structures (stars, galaxies, clusters, and superclusters) that we observe today. The structure of the CMB reflects the initial conditions from which all structure in the universe formed.

The remarkable uniformity observed in the CMB has challenged both theorists and experimentalists. Theoretically, it is difficult to understand how the structures that we observe today could form gravitationally from the extremely homogeneous initial conditions implied by the isotropy of the CMB. Theorists are forced to speculate that most of the matter density in the universe is in an exotic form that does not interact electromagnetically, and has never been observed in the laboratory (Bond et al., 1991; Vittorio et al., 1991).

Experimentally, it is challenging to measure structures in the CMB with the 1 ppm precision required to constrain the theory. This goal is particularly daunting for ground-based and balloon-borne experiments, which must observe through an unstable atmospheric foreground that is comparable in brightness to the CMB. An orbital mission, the COsmic Background Explorer (COBE), was the first to detect statistical evidence for structure in the CMB at the 10 ppm level (Smoot et al., 1992).

Space Science Reviews **74**: 151–156, 1995.

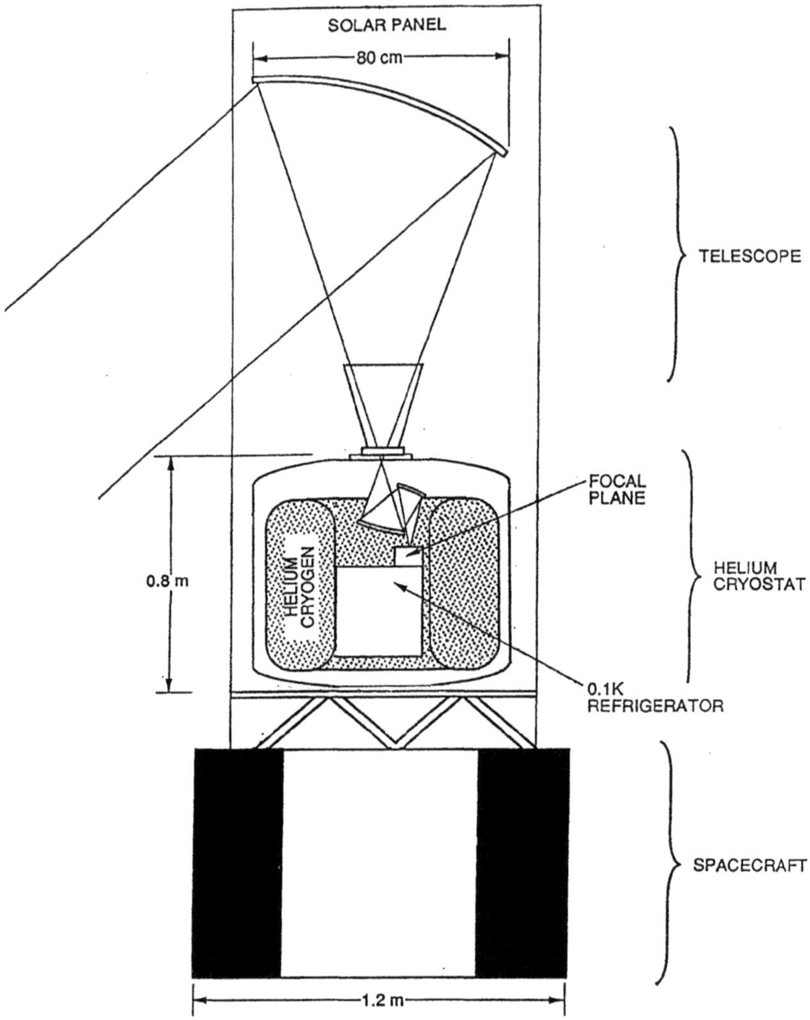

Fig. 1. A side view of the FIRE. The spin axis of the spacecraft is pointed into the page, towards the sun. The telescope and cryostat are shielded from the sun, and passively cool to < 70 K. The entire sky is surveyed in 6 months

In order to adequately constrain theory, it will be necessary to map the CMB with much higher angular resolution than COBE. In addition, future missions will require broader frequency coverage in order to accurately separate structure in the CMB from structure in the foreground emission from our galaxy, and from confusion by extragalactic sources such as radio and IR bright galaxies.

The Far IR Explorer (FIRE) is an orbital telescope that will provide an all-sky map of the CMB with 20 to 100 times the spatial resolution of COBE and with 30 times the sensitivity to anisotropy per pixel as COBE. These capabilities will allow FIRE to make a definitive measurement of the structure of CMB anisotropy down to angular scales that correspond to structures that we can map out in the present universe. It will thus provide a direct test of the theory of the formation of structure in the universe today. In addition, FIRE's broad spectral coverage will allow the galactic foregrounds to be accurately separated from the CMB, and will allow CMB anisotropy induced by Compton scattering of the CMB along the line of sight after recombination to be distinguished from primordial temperature fluctuations. Finally, FIRE's sub-mm coverage with high spatial resolution will provide an all-sky map of the most luminous objects in the universe at distant redshift.

2. Mission Concept:

FIRE is designed to be a low-cost mission that will fit into NASA's MIDEX category. Recent advances in detector technology enable an enormous improvement over COBE in sensitivity, angular resolution, and spectral coverage at relatively modest cost.

The characteristics of the mission are outlined in Table I. The FIRE telescope, shown in Figure 1, is an 80 cm off-axis parabolic dish that is passively-cooled to 70 K. A superfluid He cryostat containing re-imaging optics and a secondary cooler sits at the prime focus. An array of bolometric detectors (see Bock et al., this volume) cooled to 0.15 K provides high sensitivity at or near the diffraction limit across the entire frequency range from 50 to 1500 GHz. The bolometric detectors are operated in total power mode using an AC bridge readout technique (Wilbanks et al., 1992) that provides good stability over timescales of minutes.

FIRE will be launched into a heliocentric orbit by a Taurus XL launch vehicle. The heliocentric orbit provides several important advantages over a low earth orbit. First it allows efficient passive cooling of the primary mirror and of the outer shell of the He cryostat. Cooling the primary mirror reduces the background noise due to emission from the mirror surface. Cooling the outer shell of the cryostat dramatically increases the lifetime of the cryostat, allowing a one-year lifetime to be achieved within the strict mass and volume limits of the Taurus launch vehicle.

Second, the heliocentric orbit places the Earth at a distance of $> 10^6$ km early in the mission, eliminating the dominant source of radiation viewed by the sidelobes of the instrument. Finally, the heliocentric orbit allows a scan strategy that is extremely favorable for passive cooling of the telescope

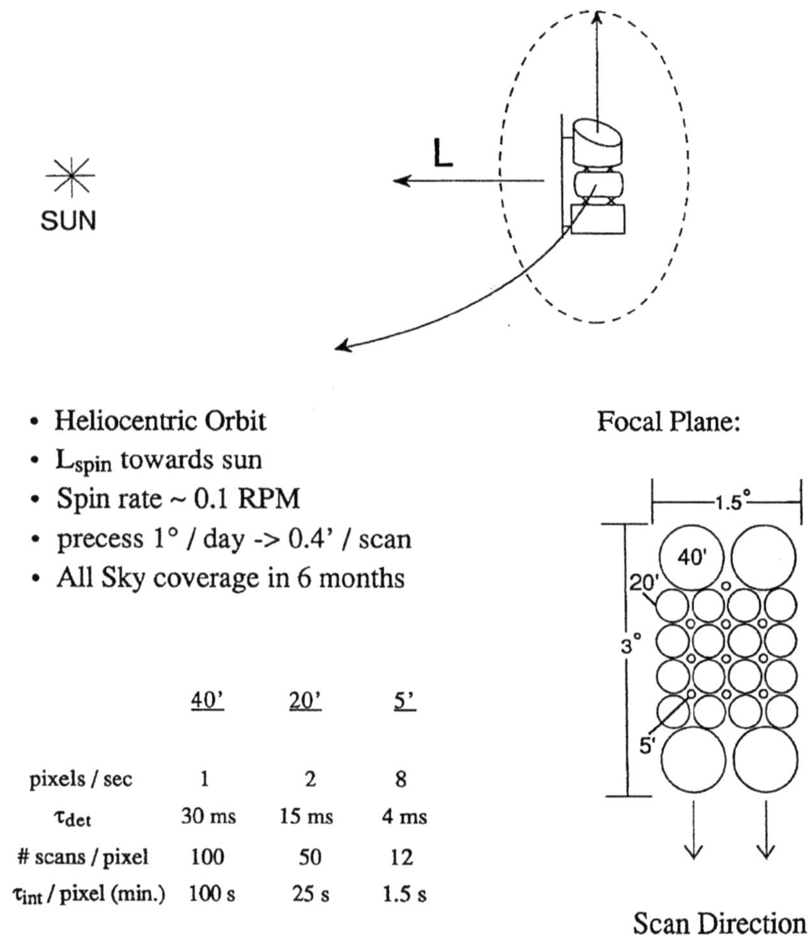

Fig. 2. One configuration of the FIRE focal plane and scan strategy. The CMB will be mapped in 4 bands between 90 and 450 GHz with 20′ resolution. The 4 × 4 array of 20′ pixels provides 4-fold redundancy in each band and 100 s of total integration time per pixel in 6 months. Additional channels provide 40′ resolution at 50 GHz and 5′ resolution above 450 GHz.

and cryostat, places modest demands on the spacecraft, and results in an efficient mapping of the entire sky in 6 months.

The scan strategy is illustrated in Figure 2. The satellite spins at a constant rate about an axis that is perpendicular to the telescope boresight and is always pointed directly towards the sun. A spin rate of 0.1 rpm is slow enough to allow for thermalization of each bolometric detector on each pixel of the sky, and fast enough to allow for long term drifts in the detector system to be accurately measured by observations of the North and South

TABLE I
FIRE Specifications

Frequency Coverage	50 GHz – 1500 GHz
Angular Resolution	5′ ($\nu > 450$ GHz) to 40′ (50 GHz)
Lifetime	1 year
Sensitivity	$\Delta T/T = 10^{-6}$/ pixel, all sky
Cost	< 75 M$
Mass	< 340 kg
Orbit	Heliocentric
Launcher	Taurus XL
Telescope	80 cm, off-axis, passively cooled
Detector Cooling	He cryostat + (ADR or Dilution Refrigerator)
Focal Plane	150 mK bolometric detectors / no moving parts

TABLE II
FIRE Sensitivity Limits

Freq (GHz)	$\Delta\theta$ (arcmin)	$\Delta T/T$ ($\times 10^{-6}$)	Δy ($\times 10^{-6}$)	ΔF_ν (mJy)
60	40	0.8	0.4	6.5
120	20	1.2	1.0	8.0
240	20	0.9	5.0	9.0
450	5	75	20	35
900	5			70
1500	5			110

1σ sensitivity per pixel in 6 months.
T = 80 K optics; $\epsilon = 4 \times 10^{-3}\sqrt{1\text{mm}/\lambda}$
80 cm primary
30% optical efficiency
$\Delta\nu/\nu = 0.5$
$NEP^2 = 2 \times (4k_B T^2 G + NEP^2_{\text{BLIP}}); G = 15Q/T_c$

ecliptic poles, which are viewed on every scan throughout the mission. The scan plane is precessed at 1 degree/day as the satellite orbits about the sun, providing an all sky map in 6 months.

The cold re-imaging optics provide a large useful field of view and vastly reduce the amount of emission from the passively cooled telescope that reaches the detector. An array of detectors in the focal plane allows redundant frequency coverage in 7 passbands and the flexibility to use the telescope near its diffraction limit in each band. For the configuration of pixels shown in Figure 2, the CMB is mapped with 20 arcminute resolution in 4 passbands, each with a total integration time per band of 100 s/pixel in 6 months.

The sensitivity of the maps produced by FIRE is indicated in Table II, in terms of fractional temperature change of the CMB, the Compton "y" parameter that describes the interaction of the CMB with hot plasma along the line of sight, and the sensitivity to point sources in mJy. At 60 to 240 GHz, the mission will reach a sensitivity of close to 1 ppm/pixel over the entire sky in 6 months. This is close to estimates of the ultimate limit determined by confusion from the galactic foreground. At higher frequencies, the increased angular resolution will provide a sensitive map of infrared point sources.

Most of the key technologies for FIRE have been or are now being developed, and will be tested on a long-duration, balloon-borne telescope (BOOMERANG – see Lange et al., in this volume) in 1995.

3. Conclusions

FIRE represents an enormous step forward in imaging the CMB. It will provide an all sky map of primordial structure in the CMB with 20 arcminute resolution and a sensitivity of 1 ppm/pixel. In addition, the large frequency coverage and high spatial resolution will provide a powerful probe of distant clusters of galaxies and of infrared luminous galaxies in the early universe. The FIRE mission uses advanced cryogenic and detector technologies to achieve these goals within the tight cost, mass, and volume envelopes of NASA's MIDEX program.

4. Acknowledgements

We gratefully acknowledge the help of the FIRE working group, especially C. Beichmann, M. Birkinshaw, N. Gautier, P. Richards, G. Rieke, and M. Werner, and of D. Elliot, L. Rosenberg, R. Rowley, and H. Schember at JPL. This work has been supported by the Office of Space Science and Instruments at JPL.

References

Bock, J.J. et al.: 1994,*this volume.*
Bond, J.R. et al.: 1991, *Phys. Rev. Lett.* **66**, 2179
Lange, A.E.et al.: 1994, *this volume.*
Smoot, G.F. et al.: 1992, *Ap.J.* **396**, L1
Vittorio, N. et al.: 1991, *Ap.J.* **372**, L1
Wilbanks, T. et al.: 1990, *IEEE Trans. Nucl. Sci.* **37**, 566

OLBERS
AN INTERPLANETARY PROBE TO STUDY VISIBLE AND INFRARED DIFFUSE BACKGROUNDS

An answer to the Call for Mission Concepts for the ESA Follow-up to Horizon 2000

F.-X. DÉSERT
Institut d'Astrophysique Spatiale
Bât. 121, Université Paris XI,
91405 Orsay Cedex France

Abstract.
The visible extragalactic background (though as yet undetected) is insufficient to explain the abundance of heavy elements in galaxies: either there should be some diffuse extragalactic light in the near infrared (from 1 to 10 μm) and/or in the far infrared ($\geq 100\,\mu$m) if dust has reprocessed the star light. We propose a new space mission to be dedicated to the search and mapping of primordial stellar light from the visible to the mid-infrared (20 μm). In this spectrum range, detectors have reached such a sensitivity that the mission should aim at being (source) photon noise limited, and not any longer background photon noise limited. For that purpose, a small passively cooled telescope with large format CCDs and CIDs could be sent beyond the zodiacal dust cloud (which is absent beyond a solar distance of about 3 AU). In that case, the only remaining foregrounds before reaching the extragalactic background, is due to the Milky Way integrated emission from stars and the diffuse galactic light due to scattering and emission by interstellar dust, which are all unavoidable. Maps of the extragalactic light could be obtained at the arcminute resolution with high signal to noise ratio. This mission is the next logical step after IRAS, COBE and ISO for the study of extragalactic IR backgrounds. It has been proposed as a possible medium-sized mission for the post-horizon 2000 ESA program that could be a piggy back of a planetary mission.

Key words: Visible and Infrared diffuse backgrounds

1. Scientific Objectives

1.1. Introduction

The measurement of the sky brightness has been a long standing problem. Wilhem Olbers (1758-1840), among others like de Chéseaux in 1744, noted in 1826 that it was paradoxical that the night sky was so dark. The solution of the paradox is still not completely settled though ground and space-based astronomy has put it on firm quantitative grounds. Elements of the solution include: interstellar dust clouds (blocking UV and visible light), finite size of the Milky Way, expansion, finite speed of light, and finite lifetime of the Universe as a whole (see Harrison, 1987). Indeed, modern instruments can easily pick up backgrounds from gamma to radio wavelengths. One aspect of the problem we are mainly concerned here with is the search for light

Space Science Reviews **74**: 157–162, 1995.

from stars in primordial galaxies. A major question of visible and infrared extragalactic astronomy is to know how and when the heavy elements like carbon, oxygen, nitrogen, etc. (that cannot have been made in the Big Bang nucleosynthesis) got made as we can see in present galaxies. The answer so far is that the visible extragalactic background (though as yet undetected) is insufficient to explain the heavy elements: either there should be some diffuse extragalactic light in the near infrared (from 1 to 10 μm) or/and in the far infrared ($\geq$ 100 μm) if dust has reprocessed the star light.

1.2. General goal of the mission

We propose a new space mission to be dedicated to the search of primordial stellar light from the visible to the mid-infrared (20 μm). In this spectrum range, detectors have reached such a sensitivity that the mission should aim at being (source) photon noise limited, and not any longer background photon noise limited. The IAU symposium Nr 139 shows the current knowledge on galactic and extragalactic backgrounds. The consequences of this goal are as follow: the airglow from the Earth atmosphere gives such a foreground that the mission has to be in space. Figure 1 shows the sky brightness observed from space at various wavelengths. The next foreground is due to interplanetary dust: below 5 μm it produces a strong scattering of solar light and above, it thermally emits a dominant foreground. We think that it is now within current technological possibilities to envision a small telescope being sent beyond the zodiacal dust cloud (which is absent beyond a solar distance of about 3 A.U.). In that case, the only remaining foregrounds before reaching the extragalactic background, are due to the Milky Way integrated emission from stars and to diffuse galactic light due to scattering and emission by interstellar dust, which are all unavoidable (continuous curve in Figure 1a). The study of galactic and extragalactic light is the prime target of this mission, but during the cruise from Earth to beyond 3 AU, the zodiacal light could also be studied in great detail for the first time. In order for the instrument to work at the largest throughput but with a sufficent spatial resolution to enable background fluctuations measurements, the angular resolution should be between 1 and 10 minutes of arc.

2. Conceptual Description

2.1. The instrument

The instrument is composed of a telescope with an estimated primary diameter of 10 to 50 centimeters. Stray ligth can be avoided by careful baffling and by using an off-axis primary mirror. A configuration with 3 mirrors like in the ground-based LITE project allows imaging of a large unaberrated field of view of several square degrees. Light is then split according to wavelength band by dichroics and/or filters and falls on various mosaics of detectors:

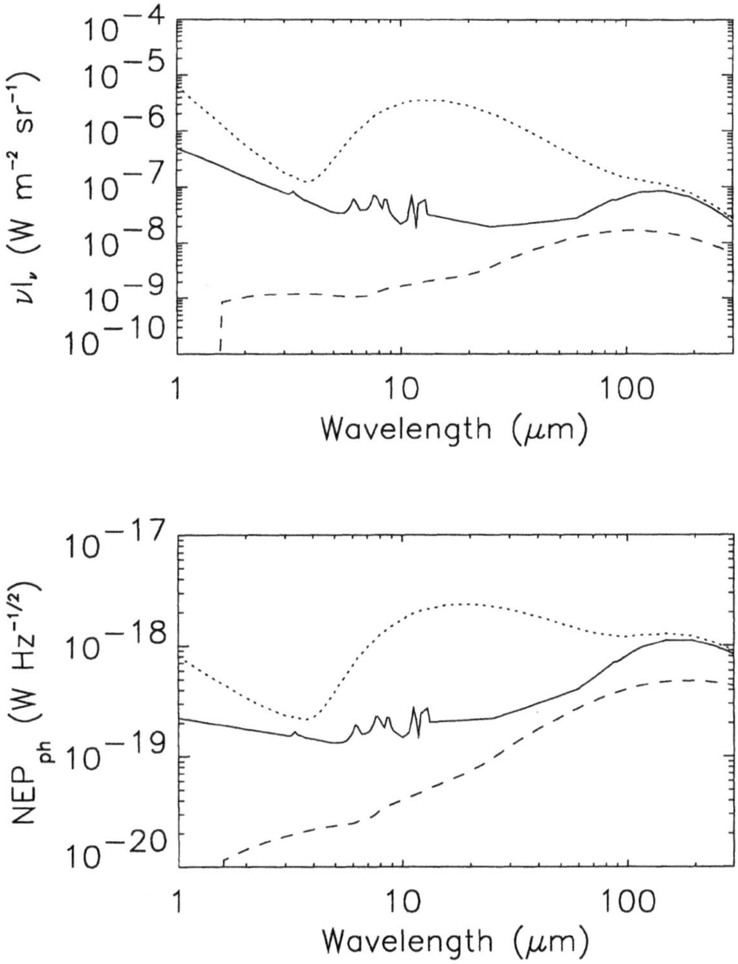

Fig. 1. a) Sky brightness at the ecliptic poles: dotted curve is the observed brightness at the distance of the Earth, continuous curve is the brightness that would be observed at a distance larger than 3 AU from the Sun, dash curve is an estimate of the sought for extragalactic background (not taken into account in the previous curves). It is expected at the 10^9 to 10^8 W m^{-2} sr^{-1} level from the visible to the mid infrared which is 10 to 100 times the expected sensitivity of the proposed instrument b) Photon noise equivalent power for an ideal detector with a spectral resolution of 4 (see Lamarre et al., this conference).

state-of-the-art detectors can have large formats (up to 512 by 512 pixels) throughout the wavelength coverage that is considered: 0.4 to about 20 μm, and which is necessary in order to subtract the galactic foregrounds and to measure the extragalactic background. One can expect that, with a careful design, passive cooling of the telescope can easily achieve 80 K just after the start of the mission, and down to 30 K or less when the distance to the Sun exceeds 3-4 AU This temperature is needed, mostly for wavelengths above 5 μm, for the telescope emission to be negligible onto the detectors. One cryocooler would allow the appropriate mid infrared detectors to work below 10 K.

2.2. The vehicle

The studies about Rosetta (ESA third cornerstone) has shown the feasability of platforms delivering large power even at large distance from the Sun and having good pointing stability. The requirement (15 seconds of arc) is met by Rosetta and future planetary missions. The proposed instrument could therefore be added to one of the future outer solar system missions and be used during most of the cruise to the planets.

2.3. Expected scientific return

The mission can be programmed in two parts: a shallow survey covering most of the accessible parts of the sky and deep surveys near the ecliptic poles. Accurate and fully calibrated maps of the sky at visible and infrared wavelengths would be the starting data for a full analysis of the contributions of the various foregrounds and backgrounds. Combined with point source catalogues made by ground-based telescopes (e.g. digitised Palomar plates and K-band surveys like DENIS), one could extract the galactic and then extragalactic components from these maps. Simple calculations show that sensitivities of 0.3 to 0.03 $\mathrm{nW\,m^{-2}\,sr^{-1}}$ can be achieved for one pixel (in typically one hour, one σ) which is 10 to 100 times less than the expected extragalactic background (and 1000 times better than IRAS at 12 μm). Detailed studies of the fluctuations of this background seem therefore possible. The mid infrared part of the instrument is essential to study the interstellar dust emission in order to obtain the extragalactic NIR background by substraction. We think that the instrument should work in a survey mode (like IRAS) from a pre-defined planning by a dedicated scientific team.

3. Technology status

This mission can be achieved with existing space-qualified off-the-shelf technology. The modern detectors: CCDs for the visible, NIR detectors (InSb...) and BIB arrays for the longer wavelengths have now a very low readout noise (less than 50 electrons) required by the mission. They are thin enough to be the least sensitive to cosmic rays and they have low dark current levels so that integration times can be of few to 20 minutes. The main limit then comes from the photon noise as estimated in figure 1b. Cryocoolers (BAe) are now available with low weight and consumption. Passive cooling of telescopes has been thoroughly studied by space agencies and should be possible at the required level.

4. Relation to existing or planned space and ground systems

The "pioneering" experience similar to what is proposed here was the IPP instrument on board Pioneer 10 (Hanner et al., 1974, Soberman et al., 1974), with a field of view of about 2 degrees and 2 channels in the blue and red. It showed that the interplanetary dust is within 2.8 AU from the Sun and measured with some accuracy the brightness of the Milky Way and put some limits on the visible extragalactic background at less than 6 $nW^{-2} sr^{-1}$. Other related projects are the IRAS then COBE (DIRBE) satellites which mapped the sky in the infrared range (resp. 1 and 40 minutes of arc of resolution). The problem of giving accurate values of the background can be pinned down to the difficulty of modelling the zodiacal emission and scattering accurately enough. Even if a subtraction method can work, the end result of these missions will be an estimate of the extragalactic background with no analysis (as we propose) of the background fluctuations. The same can be said of any measurement by HST or Hipparcos. The proposed mission is a parallel of SAMBA, a proposal of a medium-sized satellite (submitted to ESA) to measure the fluctuations of the 3 K cosmic microwave background as accurately as measured by COBE on large scales: if the experiment DIRBE on COBE gives a value of the near infrared background, the next logical step is to study its structure. The Japanese team lead by T. Matsumoto has launched rockets to make a low resolution spectrum of the NIR background light. They plan longer integration observations with the IRTS telescope on the SFU Japanese satellite (1995). They will cover one modest beam of 8 $arcmin^2$ at a time. Their rocket measurements are difficult to interpret because of the contribution of the zodiacal light. The near infrared catalogue of point sources that will be done with ISO and K-band ground based surveys (e.g. DENIS) will be used to remove their contribution to the diffuse backgrounds. Note that ISO and DENIS may not be able in their deepest surveys to observe objects individually if their redshift is large (≥ 2) but these objects may contribute substantially to the extragalactic background and its fluctuations measured by our proposed instrument (see e.g. Boughn et al., 1986).

5. Mission requirements

The requirement of the lowest possible foregrounds implies to put the instrument on a space probe on cruise to distant planets of the solar system. A lifetime of over 5 years would mean a large sky coverage. Stability and pointing can be achieved if the instrument is installed on a planetary mission of the Rosetta type. The power consumption is estimated at 120 W for the cryocoolers and 40 W for the electronics and detectors (Bradshaw et al. this conference). One should then include the consumption for the control of the platform attitude and the telemetry. Assuming a compression ratio of 5 and data transmission rate of 10 kbit/s (available with Deep Space Network), several 10-20 minutes integration large format images can be sent back to Earth. The instrument, if considered as a "piggy back" experience, should be well within a medium-sized mission financial envelope. The project still needs a complete feasability study. Yet it is a mission concept which is within current technological and financial possibilities. There are many European space laboratories that could collaborate on this project which is quite different from other space probes and satellites in that it would give original results unreachable with other concepts.

6. Acknowledgements

The author wish to thank some discussions with the group "Physics of galaxies" at IAS, L. Vigroux and D. Rouan. A science group led by John Mather at Goddard Space Flight Center is also developing these ideas in the USA. Martin Harwitt (1981) also advocated an interplanetary probe to go beyond the zodiacal cloud in order to observed diffuse infrared backgrounds, a long time ago.

References

IAU Symp. 139: 1990, S. Bowyer and C. Leinert, eds., *The Galactic and Extragalactic Background Radiation*, Kluwer:Dordrecht
Boughn, S. P., Saulson, P. R., and Uson, J. M.: 1986, *ApJ* **301**, 17
Hanner, M. S., et al.: 1974, *Jou. of Geophys. Res.* **79**, 3671
Harrison, E.: 1987, *Darkness at Night*, Harvard University Press:Cambridge Ma
Harwitt, M.: 1981, *Cosmic Discovery*, The Harvester Press:Brighton UK
Soberman, R. K., et al.: 1974, *Jou. of Geophys. Res.* **79**, 3685

HOW TO EVIDENCE PRIMITIVE LIFE ON AN EXO-PLANET? – THE DARWIN PROJECT –

A. LÉGER and J.-L. PUGET
Institut d'Astrophysique Spatiale
Bât 121, Université Paris XI,
91405 Orsay Cedex France

and

J.M. MARIOTTI, D. ROUAN and J. SCHNEIDER
Observatoire de Paris
F-92195 Meudon, France

Abstract. We report the concept of an IR observatory, with interferometric rejection, to search for Primitive Life on an extra-Solar Planet which has been proposed to the European Space Agency.

Key words: Bioastronomy

1. Scientific objectives

1.1. Motivation for the search of primitive life

The search for life in the universe will be a central item in the scientific activity of the 21st century. The TOPS report (1992) states: "Much of religion and philosophy has focused on attempts to understand how we and our world came to be". We "ask about the prevalence of planetary systems throughout the universe (...) and the likelihood that other planets have given birth to life, even advanced and intelligent forms of lives. *We live in a remarkable time,* when human beings (...) have attained the possibility of finding *the first real answers* to some of these most meaningful questions".

As financial resources are scarce, it is important to distinguish which are the key steps. Current opinion states that we are to find exoplanets in the next 10–15 years (e.g. by astrometry with the European VLTI). The knowledge of the frequency of appearance of primitive life on habitable planets would be a capital breakthrough, whereas the only detection of Jupiter like exoplanets would not really answer our basic quests. The DARWIN concept is one of the very few possibilities we have to address this question.

1.2. How to find the evidence for it? Why a space mission?

The presence of abundant O_2, or O_3, in the atmosphere of an exoplanet has been shown to be a criterion for photosynthetic activity (Lovelock, 1975). The main detectable signatures of these species are in the visible (O_2: 760, 680 nm), in the IR (O_3: 9.6 μm) and in the radio (O_2: 5 mm). The latter

Space Science Reviews **74**: 163–169, 1995.

is not attractive because of the λ^{-2} dependence of the number of emitted photons for a 300 K thermal source.

The main challenge when searching for these signatures in a planet atmosphere is the presence of the nearby, much brighter, star. In the Sun/Earth case the luminosity ratio is 7×10^9 in the visible and 10^7 at $10\,\mu$m, pointing out the advantage of observing in the IR (Bracewell, 1978). After the starlight is rejected as much as possible, the signatures, if present, appear as dips in the spectrum. Considering the *quantum noise* alone, the feasibility of the detection can be evaluated by calculating the integration time (t) required for a $5\,\sigma$ detection of a 50% deep spectral band :

- O_2 band in the visible (760 nm...)
  With large ground based interferometers (e.g. VLTI) in the visible, the maximum rejection factor (ρ) of the starlight with respect to the planet emission, if ever possible, will be 10^2. With a telescope–detector yield of 15% :
  $$t = 3 \times 10^5 \left(\frac{L}{10^{-2} L_\odot}\right) \left(\frac{\rho}{10^2}\right)^{-1} \left(\frac{R_{\rm pl}}{R_\odot}\right)^{-4} \left(\frac{D}{4\,\rm pc}\right)^2 \text{ hours.}$$
  Even for nearby M stars, this is not possible.
- O_2 band (760 nm...) during a planet transit
  When a planet transits in front of a star (Schneider & Chevreton, 1990) a small fraction of the star flux travels through the planet atmosphere and is specifically absorbed. For a M star system, this occurs during 30 hours per year but the number of years needed for the band detection is prohibitive ($\simeq 10^3$, Schneider et al., 1994).
- O_3 band in the IR (9.6 μm)
  A null interferometer in space can reject the starlight (black fringe) and transmit the planet's light (bright fringe) if it can resolve the system. For a planet at the suitable distance from the star in order to be at 300 K (Kasting et al., 1993), the condition to separate the star from the planet is :
  $$D < 20 \left(\frac{L}{L_\odot}\right)^{\frac{1}{2}} \left(\frac{l}{20\,\rm m}\right) \text{ parsec}$$
  D : distance of the system, L : star luminosity, l : interferometer base. Much higher rejection factors than on ground are possible and, assuming that all other sources of noise are at levels below that due to stellar light –as explained below– the integration time becomes:
  $$t = 34 \left(f\frac{L}{L_\odot}\right) \left(\frac{D}{6.3\,\rm pc}\right)^2 \left(\frac{\rho}{5\,10^4}\right)^{-1} \left(\frac{R_{\rm pl}}{R_\oplus}\right)^{-4} \left(\frac{Res}{20}\right) \left(\frac{\Phi_2}{1.3\,\rm m}\right)^{-2} \text{ hours}$$
  f: depends upon the star temperature (0.7 - 5.5 for stars of spectral type F–M), $R_{\rm pl}$: planet radius, Res: spectrometer resolution , Φ_2: diameter

of the twin telescopes. Note that there are about 100 stars closer to us than 6.3 pc.

Clearly, among these possibilities, the last is the *only one feasible.* The ability of a space mission to accommodate a cold telescope and to allow high interferometric contrasts makes it possible, although difficult. The detection cannot be done from the ground.

2. Conceptual description

The concept is a null interferometer associated with an IR spectrometer. It has two versions: (i) the "minimum" DARWIN mission, made of 2 × 1.3 m telescopes with a 10 m base, and (ii) the "full" DARWIN mission, with 4 × 0.9 m telescopes in a cross pattern, with 30 m bases. The spectral range is 5 - 25 μm or broader. At 9.6 μm, the number of photons received per hour in one spectral element from the planet is:

$$N_{\rm pl} = 9.6 \times 10^2 \left(\frac{D}{6.3\,{\rm pc}}\right)^{-2} \left(\frac{\Phi_2}{1.3\,{\rm m}}\right) \left(\frac{Res}{20}\right)^{-1} \left(\frac{R_{\rm pl}}{R_\oplus}\right)^2 \ {\rm photons.\,hours}^{-1}$$

The problem is to extract such a tiny flux out of strong ones. Sources of noise and spurious signals are discussed below.

2.1. Starlight leaks

At 9.6 μm, the ratio of the residual flux from the star to that of the planet is :

$$\frac{F_{*\rm res}}{F_{\rm pl}} = 200 \left(f \frac{L}{L_\odot}\right) \left(\frac{\rho}{5 \times 10^4}\right)^{-1} \left(\frac{R_{\rm pl}}{R_\oplus}\right)^{-2}$$

It critically depends upon the rejection factor ρ. The factors limiting ρ are :

- Optics quality and micro-roughness: the amount of light scattered to the dark field by these optical defects is proportional to the variance of the optical path difference errors. Achieving an optical quality of $\lambda_{\rm vis}/100$ in the visible (0.5 μm) will limit the scattered light at 10 μm to 10^{-5}. Other sources of stray light have to be smaller.
- Optical path cancellation : an error δ in optical path adjustment will produce light leak. $\delta = \lambda_{\rm IR}/2000$ will result in a 10^{-5} level of light intensity. But if the control of the adjustment is done in visible light (0.5 μm), where photons are numerous, the requirement drops to $\lambda_{\rm vis}/100$. A further requirement is that the optical path difference must be achromatic in the IR.
- Guiding : the cancellation of the fields in the interferometer is perfect only if the object is on axis. If the interferometer axis is displaced by only 1/1600 of the angular size of the Airy disks formed by each telescope, the light leak is 10^{-5}. However, if the tracking is performed at 0.5 μm,

the requirement is reduced to 1/80 of the Airy disk size (1.3 mas, for a 1 m telescope).

- Star size : although the central star is not resolved by the interferometer, it is not a perfectly coherent source, resulting in a light intensity proportional to $R^2\alpha^2$ where α is the angular size of the star, and R the angular resolution of the interferometer ($R = l/\lambda$).

The first three limitations are technological in nature and can be each kept within the quoted limits with available technologies used at best. Their total contribution will hence be less than 2×10^{-5}. The last one is more fundamental and results in a severe trade-off for the resolution. However, Angel (1989) has proposed an optical arrangement involving 4 telescopes in a diamond pattern which yields light leaks proportional to $R^4\alpha^4$, keeping these leaks to levels much smaller than 10^{-5} at the expense of an increased complexity. This leads to the "full" DARWIN concept. Search for other clever optical concept should be pursued but it is comforting to already have one concept which fulfils the requirements.

2.2. Zodiacal light likely to be associated to the planetary system

At 10 μm, the solar integrated zodiacal light (ZL) luminosity is 3×10^2 larger than the earth luminosity. If the distant planetary system is similar to ours, its ZL should not generate too much noise because it would be of the order of the star light leak but rejected by the interferometer because it is centrally peaked. However, it probably carries the silicate band at 9.7 μm in emission which is close to the O_3 line to be detected. Using the modulation of the planet signal during its orbital motion will be needed to properly separate them.

2.3. Cirrus background

According to our present understanding of the physics of the mid infrared emission from cirrus clouds found by IRAS, this galactic background gives a mean flux at 9.6 μm :

$$N_{\mathrm{cir}} = 1.2 \times 10^4 (Res/20)^{-1} \text{ photons.hour}^{-1}.$$

The requirement that the corresponding integration time is less than 30 hours limits the distance of observable planets to $D < 11.5$ pc.

2.4. Solar system Zodiacal emission

From an Earth orbit, its flux,

$$N_{ZL} = 3.0 \times 10^6 (Res/20)^{-1} \text{ photons.hour}^{-1},$$

is *totally prohibitive.*

To avoid this difficulty the mission has to go further out from the Sun than the Earth orbit. The ZL background becomes weaker than the Cirrus one at $d > 3.5$ AU (Jupiter is at 5.2 AU).

2.5. Telescope temperature

A thermal emission by the mirrors inferior to the star light leaks at 9.6 μm, but also at 15 μm (CO_2 band) implies $T < 30$ K for a comfortable detection (up to 17.5 μm). This should be possible with passive cooling as shown by the studies done for the EDISON project, taking into account the distance to the Sun of more than 3.5 AU

2.6. Detector noise

Presently available Si:As BIBs individual detectors have already a dark current comparable to the cirrus background (3 $e^- s^{-1}$ at 10 K) in the 5-28 μm range. Provided the detectors are cooled to stable temperatures around 4 K, they will have the required low noise and stability.

3. Scientific return

3.1. Telluric planet detection

The presently planned ground based searches will not find earth size planets. The DARWIN concept can detect them spectroscopically, thanks to the CO_2 15 μm band. The band is expected whether life is present or not at their surface and is observed in Venus, Earth, and Mars spectra.

3.2. Photosynthesis activity

The mission can detect the presence or absence of photosynthetic activity on large scale on the terrestrial planets it will find. This detection does not depend on the specific molecule that reduces CO_2 using photons contrary to an experiment that would search for the chlorophyll spectral signature in the visible. The number of star systems in which the minimum DARWIN mission (2×1.3 m telescopes, $l = 10$ m) can give these answers is 35 and with the full DARWIN mission (4×0.9 m telescopes, $l = 30$ m) it increases to 110 in a 4 years mission (assuming 2 years of integration). Most of them are G stars (55%) or K stars (25%) which are good candidates in term of the "habitability"of their telluric planets.

3.3. Important by-products

Although dedicated to bioastronomy, the DARWIN concept would have many important by-products. Potential programmes, using the coronographic capability of the instrument, are:

- planetary science of non terrestrial planets, with spectroscopy
- studies of the central part of young stellar disks/circumstellar envelopes

– narrow line region of AGN, QSOs...

There is also the astrophysics that can be done with a 2.5 m^2 collecting area IR telescope without the disturbance of the zodiacal light, during a long mission, although the instrument is not optimised for this purpose (e.g. small field of view).

4. Technology status and requirements

A key point is the achievement of the high rejection factor as described above, it should be possible with existing technology (e.g. the optical path length equalising that planned for the VLTI) but requires developments. The deployable structure of the interferometer and the active cooling system also requires developments.

Technology heritage :

- solar panels, telemetry : ROSETTA
- passive cooling: studies for EDISON and FIRST
- IR detectors: ISO and SIRTF developments
- pointing: HST
- detector cooling: active coolers developed for FIRST
- deployment mechanism: developments for FIRST in the deployable version.

5. Relation to existing or planned space and ground systems

The best candidates to be observed would be those for which giant planets have been found by astrometric/Doppler studies from the ground. Such studies are likely to give results in less than 15 years and the DARWIN mission could be made the next step in the search for planets. For that to be possible developments and studies should be undertaken now.

6. Summary of the main mission requirements

These are

- **orbit:** more than 3.5 AU from the sun, away from solar system planets.
- **lifetime:** typically 10 years.
- **operational strategy:** autonomous.
- **stability and pointing:** 3 axis stabilisation; 1.3×10^{-3} arcsec in the focal plane achieved by locking on the star itself in the visible.
- **power:** to be determined.
- **data rate:** low, compatible with existing ones for planetary missions.
- **readiness:** major development needed: deployable structure in space, balance between the intensity and optical paths from the different telescopes.

– **cost:** cornerstone class.

7. Conclusion

The mission proposed addresses one of the most fundamental problems that science can deal with. Most of the basic technology needed for it has been (or is being) developed for other missions and should be available in the post Horizon 2000 era. Nevertheless the concept is a very ambitious one and will require careful studies and some development of new space technology. It will require a thorough system analysis to make sure that the fundamental limits we aim at can be achieved. From that point of view it fits well to the cornerstone concept as developed in ESA's scientific program.

References

TOPS: *Towards Other Planetary Systems*: 1992, B.F. Burke ed., NASA.
Lovelock J.E.: 1975, *Proc. R. Soc. Land. B.* **189**, 167
Bracewell R.N.: 1978, *Nature* **274**, 780
Schneider J., Chevreton M.: 1990, *A&A* **232**, 251
chneider J., Bézard B. and Léger A.: 1993, *in preparation.*
Kasting J.F., Whitmire D.P. Reynolds R.T.: 1993, *Icarus* **101**, 108
Angels R.:1989, in P. Bély, C. Burrows and G. Illingworth eds., *The Next Generation Space Telescope*, STScI:Baltimore, 81

THE NEAR INFRARED HIGH RESOLUTION IMAGING CAMERA PROGRAM OF BEIJING OBSERVATORY

JINGHAO SUN
MPI fur Astronomie, Heidelberg, Germany
and
Beijing Astronomical Observatory, Beijing, China

Abstract. This paper introduces the program of an adaptive optics system using an infrared camera for the near infrared observations based on the 2.16 m telescope of Beijing Observatory. This system consists of 3 parts: (1), the 2.16 m telescope; (2), the adaptive optics system that will be mounted at the coudé focus on an optical table. It will be used to remove the effect of atmospheric turbulence on the imaging observations; (3), the infrared camera with a 512×512 PtSi IR detector array.

1. The 2.16 m Telescope

The largest telescope made in China was installed at Xinglong station of Beijing Observatory. The clear aperture of the primary mirror is 2.16 m and it has an f-ratio of 3 (f/3). There are three foci in this telescope: Cassegrain (f/9), Coudé (f/45) and prime foci.

2. Adaptive Optics System

An adoptive optics system will be mounted at the coudé focus on an optical table. It will be used to remove the effect of atmospheric turbulence on imaging observations and to improve the spatial resolution. A 21-elements adaptive optics system will be used to fit the 2.16 m telescope for near infrared observations. The optical layout is shown in Figure 1.

M_0 is a mirror which reflects light from telescope to the AO system. M_1 is a spherical mirror used as a collimator. The light beam size will be fit to the size of the wavefront corrector. The latter consists of two tip-tilt mirrors and a 21-elements deformable mirror (DM) driven by piezocrystals. The first tip-tilt mirror (TM1) corrects the low frequency (2.5 Hz) and large amplitude image motions caused by atmospheric turbulence and tracking errors from the telescope. The control signal of TM1 comes from an image Tube+CCD system (I-CCD) and a position error detection system. The second tip-tilt mirror(TM2) corrects the high frequency (2.5-10 Hz) and small amplitude image motions. The deformable mirror (DM) corrects the high-order aberrations caused by the atmosphere.

TM2 and DM are driven by a control signal from a wavefront detector and processor. A dichroic mirror S_2 splits the light into two beams, one optical and the infrared. The optical beam is used for wavefront detection. In the system, the wavefront detector is a photon counting dynamic differential

Space Science Reviews **74**: 171–173, 1995.

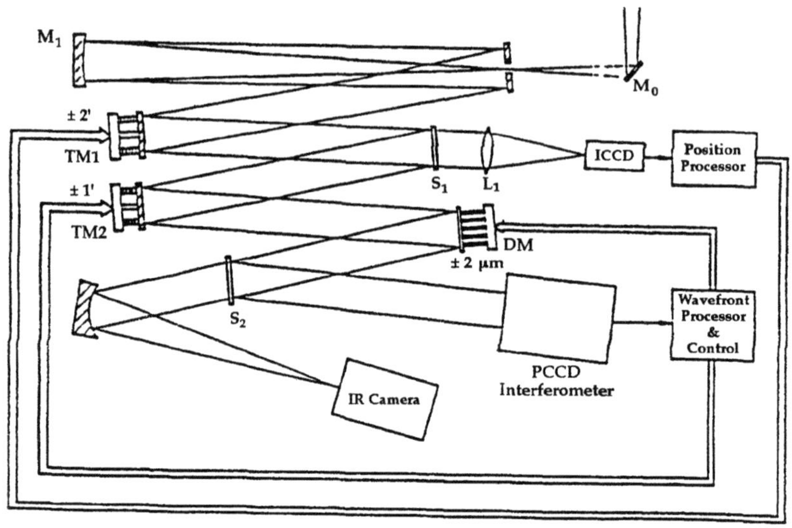

Fig. 1. Optical layout of the Adaptive Optics system.

interferometer. It can measure the wavefront tilt at each element in the direction of X and Y. The wavefront processor gives a control signal to drive the TM2 in two directions and drives piezocrystal elements in the DM. The total number of deformable elements is 21. In the X and Y directions there are 32 sub-apertures to detect the wavefront deformations. The size of each sub-aperture is about 40 cm. It is fit with the size of r_0 at $2\,\mu$m.

The AO system works as a loop-locked control system. It can correct the deformations of the wavefront due to atmospheric turbulence and improve the spatial resolution of astronomical near infrared imaging observations. It is hoped to reach the diffraction limit, i.e. for the 2.16 m telescope, about 0.3 arcsec.

3. The PtSi Infrared Camera

The 512×512 PtSi IR CSD (Charge Sweep Device) which was developed by Mitsubishi Electric Company of Japan is selected for the detector of the camera system. The PtSi Schottky-Barrier diode array does not have a high quantum efficiency but has an excellent uniformity and stability, a large format size and a low readout noise. In this case the PtSi image sensor is more effective than a high quantum efficiency detector with small format.

This camera system is shown in Figure 2. The total system contains: the camera dewar (Flank Low HDL-8), the PtSi IR CDS is cooled by solid

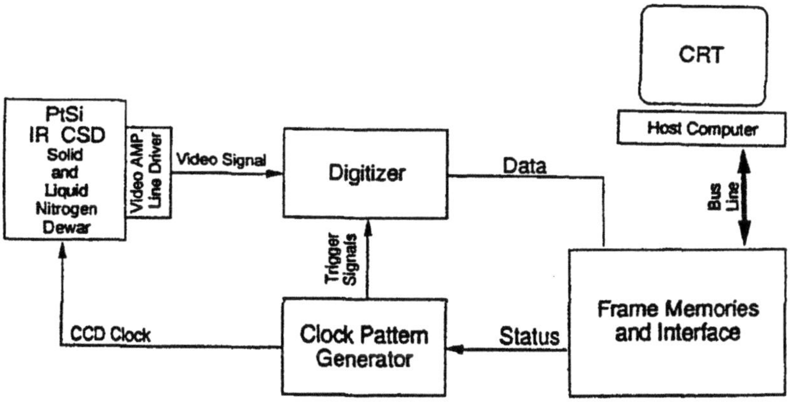

Fig. 2. The block diagram of the PtSi IR Camera System.

nitrogen down to 52 K; the preamplifier and A/D converter (ADC 4322, 16 bits, 2 MHz); the clock pattern generator (CPG); the frame memory, and the host Computer (Sun Workstation).

The wavelength range is 1-2.5 μm with wide band filters (J, H, K) and narrow band filters. The Optical Focus Reducer for the cassegrain focus is f/3 (0.6″/pixel; 5.0′ × 6.0′) and f/6 (0.3″/pixel; 2.5′ × 3.0′) without adaptive optics system.

Acknowledgements

This program is a collaboration between the National Astronomical Observatory of Japan, Beijing Astronomical Observatory and the Institute of Optics and Electronics of Chengdu, China. The Max-Planck-Institut für Astronomie of Germany has given us much help towards this collaboration. The system is going well now. We hope to be completed in Autumn, 1994.

References

Su Ding-qiang, Zhou Bi-fang et al.: 1990, *Science in China (Series A)* **Vol. 33**, **No.4**, 454

Ueno, M., Ito, M., et al.: 1992, 'Infrared Technology XVIII', *Proceedings of SPIE*, **Vol. 1762**, 423

THE EFFECT OF REALISTIC SURFACE PROPERTIES ON LOW TEMPERATURE SPACE OBSERVATORIES

ROB BLAKE and BARRIE W. JONES
Physics Department, The Open University
Milton Keynes MK7 6AA, UK

Abstract. We highlight the effect on space-telescope temperatures of the *directionality* of the radiative properties of materials, by showing results from a Monte-Carlo simulation of telescope cooling. The need for further measurements of directional properties is stressed.

Key words: space telescopes – infrared telescopes – Monte-Carlo method – bi-directional reflectance – directional emittance – directional absorptance – telescope cooling

1. Overview

For infrared telescopes it is important to maximize radiative cooling, because if the equilibrium temperature of the telescope rises then:

- long wavelength sensitivity is lost in radiatively cooled telescopes
- mission lifetime is shortened in cryogenically cooled telescopes.

In order to reduce the equilibrium temperatures, the materials of the telescope have to be selected carefully, and in order to calculate the cooling rates and the equilibrium temperatures the materials must be well characterized. In particular, we need to know, for each material:

- the directional-spectral emittance $\epsilon'_\lambda(\theta, \phi, \lambda, T)$
- the directional-spectral absorptance $\alpha'_\lambda(\theta, \phi, \lambda, T)$
- the bi-directional-spectral reflectance $\rho''_\lambda(\theta_\mathrm{i}, \phi_\mathrm{i}, \theta_\mathrm{r}, \phi_\mathrm{r}, \lambda, T)$

where T is the temperature of the material, λ is the wavelength, and the various angles θ and ϕ are defined in Figure 1.

A common simplification is to assume that materials are diffuse in emittance and absorptance, and either diffuse or specular in reflectance. **Part of our work aims to investigate whether it is important to use more realistic directional properties.** This links to the rest of our work

- to specify how directional properties can be "tuned", to maximise cooling
- to measure the directional properties, there being a huge lack of data even for fairly common materials.

2. A Monte-Carlo Investigation of the Importance of Directional Properties

To investigate the importance of using realistic directional properties we are developing a Monte-Carlo simulation (Single & Howell, 1992, see section 11–

Space Science Reviews **74**: 175–179, 1995.

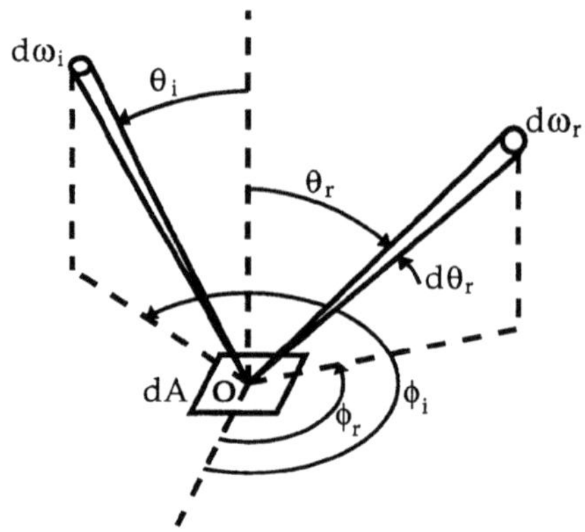

Fig. 1. The four directional angles in bi-directional reflectance. Radiation is incident on an element of surface, area dA at O; $d\omega_i$ is an element of solid angle centred on the incident beam, and $d\omega_r$ that centred on the reflected beam.

6) of radiative energy transfer in space telescopes, for the case of a cylindrical telescope tube shielded from the Sun by a flat plate, as in a simplified version of the proposed EDISON space telescope.

For the side of the flat plate facing the Sun we use the properties of silver-coated Teflon with $\alpha' = 0.08$ at solar wavelengths at $\theta = 0$, and the hemispherically averaged emittance $\epsilon_h = 0.79$ at infrared wavelengths. The cylinder is open at one end, where $\epsilon_h = 0.80$.

For the side of the plate facing the cylinder, and for the cylinder itself we have investigated contrasting cases of directional properties

- emittance and absorptance: **either** diffuse, **or** with the directional properties of smooth gold at infrared wavelengths, **or** "bizarre"
- bi-directional reflectance: **either** diffuse, **or** specular.

Figures 2 and 3 contrast these properties, the emittance and absorptance being represented by the emitted power. Note that the hemispherical power is the same in all cases, and corresponds to gold in the infrared, with $\epsilon_h = 0.03$. Note also that the surfaces are assumed (for now) to be spectrally grey.

Figure 4 shows the temperature behaviour in various cases. It is clear that the temperature of the cylinder is significantly sensitive to the directionality of the radiative properties, in that the change in the directional behaviour of the emittance, and hence the absorptance in Figure 2 is not very large. Therefore, it can be important to use realistic directional properties, and not be content with simplifying assumptions such as diffuse or specular behaviour. The ratio, cylinder-diameter:cylinder-to-plate spacing is also important.

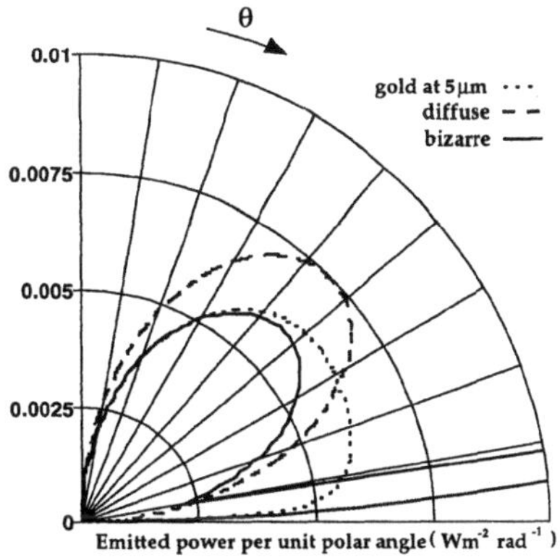

Fig. 2. The power emitted at $T = 46$ K, per unit area per radian in θ, integrated around ϕ. For the cases considered there is no dependency on ϕ. Note that the hemispherical power is the same in all cases.

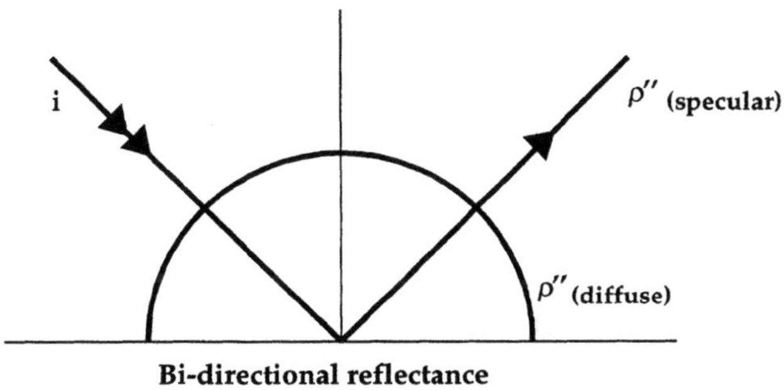

Fig. 3. The bi-directional reflectance (arbitrary units) for a diffuse surface and for a specular surface, where i denotes the incident beam and ρ'' the reflected radiation. We assume no dependency on ϕ nor on incident angle. In the simulation the infrared specular reflectance of gold is used, the diffuse case integrating to equal this value.

It is important to realise that it is possible to adjust directional properties by depositing thin films on a surface, or by texturing the surface. This would enable optimum directionality to be approached in a specific geometry of telescope. Such optima can be specified through Monte-Carlo modelling. Figure 2 shows the emitted power from a (as yet) hypothetical material with an enhanced emission at large θ. If such a surface is applied to the surface of the plane facing the cylinder, and to the cylinder, then significant extra cooling results (Figure 4).

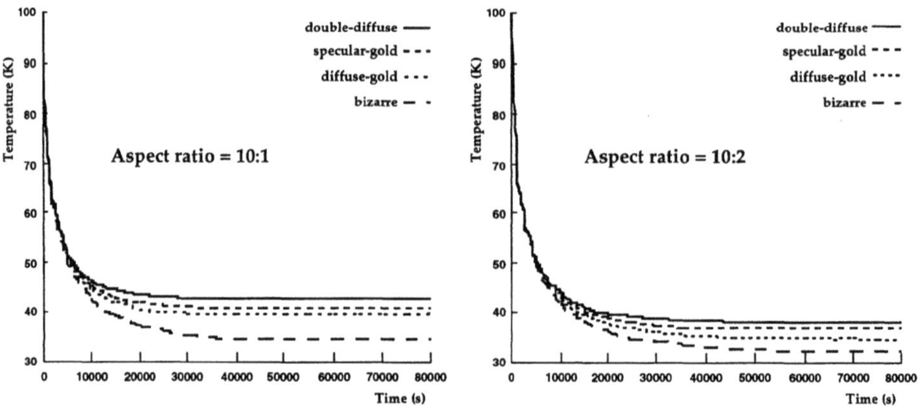

Fig. 4. Cylinder temperatures. Results from a Monte-Carlo simulation of the temperature of a cylinder, shielded from the Sun by a flat plate. The plate is square, with a side equal to the cylinder diameter, which equals its height, and 0.1 or 0.2 times this dimension equals the minimum spacing between cylinder and plate – i.e. ratios of 10:1 or 10:2. The heat capacity of the plate is 10^5 J K^{-1}, and of the cylinder is 5×10^5 J K^{-1}. The temperature curves correspond to surfaces with the following properties from Figures 2 and 3 (those for the plate are for the side facing the cylinder)

- double-diffuse: plate + cylinder diffuse in emission/absorption and in reflection
- gold-diffuse: plate + cylinder as gold in emission/absorption; diffuse reflection
- gold-specular: plate + cylinder as gold in emission/absorption; specular reflection
- bizarre: plate + cylinder bizarre in emission/absorption: specular reflection.

The plate temperatures in all these cases are close to 220 K.

Further modelling will be done over the next two years, including spectrally selective (non-grey) surfaces, and surfaces with "bizarre", but achievable directional properties.

3. The measurement of directional properties

There is a huge lack of data on the directional radiative properties of materials (Siegel & Howell, 1992, see Chapter 5). Therefore, we are re-developing equipment that will enable us over the next two years to carry out further measurements at various material temperatures of:

- the bi-directional-spectral reflectance at a selection of infrared wavelengths
- the directional-spectral emittance at a variety of visible and infrared wavelengths, and the broad-band directional emittance over much of the infrared.

From either of these measurements the directional absorptance can be obtained, and from the reflectance we can also obtain the directional emittance.

Theoretical models will be used with experimental data to obtain values at temperatures, and wavelengths, at which there are no experimental data.

We intend to carry out measurements on materials when they are fresh, and when they have been artificially aged in a space-like environment.

References

Siegel, R., Howell, J.R.: 1992, *Thermal Radiation Heat Transfer*, Hemisphere: London.

THE MANAGEMENT OF STRAY RADIATION ISSUES IN SPACE OPTICAL SYSTEMS

STEPHEN M. POMPEA*
S. M. Pompea & Associates, 1321 East Tenth Street, Tucson, Arizona 85719-5808, USA

Abstract. The performance of infrared and submillimeter systems can be severely degraded by stray light. Stray light includes off-axis radiation, system diffraction and scattering effects, and thermal self-emission. The purpose of this paper is to identify several keys to preventing system degradation due to stray radiation. The first key is to apply stray light design rules and analysis techniques early in the program before the design is finalized. A systems level analysis using stray light analysis software is often necessary in order to identify more subtle problems and to assess the magnitude of their effect on system performance. Another key is to address contamination control and the choice of surface coatings early in the program. The management of stray radiation issues is extremely cost-effective, if begun early in the program, and can reduce later schedule hardships.

Key words: Space optics – Telescope design – Instrument design – Instrumentation – Stray light – Black surfaces – Baffles – Contamination – Scattering – Bidirectional reflectance distribution function – Point source normalized irradiance transmittance.

1. Introduction: What is Stray Light?

This paper will serve as a brief introduction to the field of stray light (or stray radiation) in infrared and submillimeter systems. It is a field that touches upon and involves many subspecialties of optics. Stray light is a systems level design and fabrication issue. For stray light reduction to be successful, the experience of systems engineers, coating experts, painting technicians, machinists, thermal engineers, and others must be incorporated into the design and construction of the space instrument. Often, this wealth of experience is represented or communicated to the instrument science team and system designers by the stray light analyst. This paper is an outline of current thinking in this area and some of the tools that are used. The key challenge today is not the technical aspects of stray light control, but rather using the technical knowledge early enough in the program to do some good. This management issue is at the heart of the message of this paper.

Most of this paper's technical content is central to good optical, optomechanical, and systems engineering practice. However, recent experience indicates that the effort to prevent stray light problems in space optical systems is losing ground, perhaps because of a sense of complacency. Instrument and telescope designers have led the revolt to eliminate the occurrence of the most serious stray light design problems. However, the true revolution, which consists of applying the principles of stray light control in a systematic

* Adjunct Associate Astronomer, Steward Observatory, University of Arizona, Tucson, Arizona 85721 Telephone (602) 792-2366, Fax (602) 622-2122

Space Science Reviews **74**: 181–193, 1995.

way to every system, is still in the future. Consequently, many systems still suffer serious performance degradations that would have been prevented if a more systematic approach had been used.

What exactly is stray light and how does it affect infrared and submillimeter systems? In general, stray radiation is any radiation that degrades the performance of the system (i.e., degrades the signal to noise ratio). A useful dichotomy is to consider that two forms of types of electromagnetic radiation are propagating through the system – the "wanted" part and the "unwanted" part. The wanted radiation comes from the objects of interest (the target of the observations) and is often imaged on the detector or coherently detected. The wanted radiation is controlled by the "idealized" optical system, which contains the optical elements that can be described in a simple optical design code. The ensemble of elements defined in the design code is often a simplified and highly idealized subset of the finished or complete optical system. Many mechanical and structural components that interact with the next form of radiation are missing.

There is also "unwanted" radiation which is propagated and generated by the actual optical system. This can include radiation from an unwanted source (such as sources outside of the field-of-view) or radiation from thermally emissive objects inside or outside of the optical system. It includes radiation that is diffracted, scattered due to microroughness and contamination of optical surfaces, or scattered from inhomogeneities in transmissive elements.

All radiation (whether scattered, diffracted, or self-emitted by the system) reaching the focal plane affects the system performance level. This leads to the premise that the paths of the "unwanted" radiation need as much consideration as the paths of "wanted" radiation through the lenses and mirrors. What are the paths that "unwanted" radiation can take through a system? A cassegrain system provides a good illustration. Many space infrared and submillimeter systems are designed in this configuration because of its compact, symmetric geometry.

2. The Seven Deadly Sins of Cassegrain Systems

A Cassegrain type system with an aperture stop at the primary provides a good illustration of the kinds of stray light paths and problems that can be of significance. In a classic paper, Breault and Milsom (1992) describe seven distinct stray light paths of such a system. Some paths require a reflection off of the secondary and one path requires a reflection off of both the primary and secondary. The baffle surfaces that can be seen from the focal plane of such a telescope are key surfaces for consideration of stray light paths. These surfaces are:

– The inside of the inner secondary baffle seen in reflection.

- The inside of the inner conical baffle.
- The outside of the inner conical baffle seen in reflection.
- The aperture stop seen in reflection by the secondary.
- The inside of the inner conical baffle seen in reflection.
- The inside of the inner conical baffle seen directly.
- The inside of the inner secondary baffle seen in reflection by the secondary and primary mirror.

These paths generally are present on well-baffled cassegrain systems. There are also a number of other stray light paths that can occur because a baffle surface was misplaced, or because the system was inadequately baffled. Notice that all of these paths are described from the viewpoint of the detector.

There are a number of different strategies that can be used to block or remove these paths. One common approach is to shift the stop of the system to the secondary. This has a number of very desirable effects in eliminating views of surfaces that may be illuminated by off-axis radiation. The inclusion of a Lyot stop is also very helpful in this regard. All of the paths listed above can usually be removed or minimized through careful design work.

3. Concepts for Stray Light Reduction

Many designers begin thinking about stray light from the point of view of the source of stray radiation, outside of the optical system. They consider first the off-axis angles of the sun and earth, for example. The proper place to begin is at the end of the system, at the detector. Only the paths that reach the detector need to considered. The most basic concept for stray light reduction can be illustrated in Figure 1 which shows radiation propagating through an idealized system. The stray light analyst makes a paradigm shift to consider which objects are responsible for radiation reaching the detector rather than be concerned about the possible off-axis sources of stray radiation.

The goal of this shift in thinking is to identify critical objects, i.e., objects seen from the detector plane. The critical objects must then be moved or blocked. Next, the illuminated objects which can contribute power to the critical objects are identified. The paths that send power through intervening objects to the critical objects are then identified. These paths are blocked or objects are moved to minimize the transfer of power to the critical objects.

Keeping in mind the viewpoint of the detector, a number of areas of the system deserve special attention. Only experience or a series of trade studies can properly address many of these issues. This list is not exhaustive but does describe some areas of particular importance:

- Optimal placement of aperture and field stops
- Use of Lyot stops

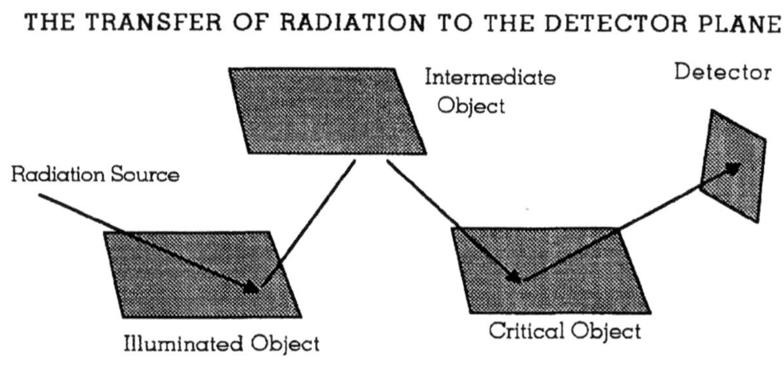

Fig. 1. Radiation is transferred from the source to the detector plane via intermediate objects. If illuminated objects are also critical objects (i.e. able to be seen from the detector) then the performance of the system will be severely degraded.

- Baffle and vane placement, and vane tip and vane cavity design
- Specular vs. diffuse surfaces for stray light control
- Strut design and coronagraphic mask design
- Reentrant spectra and order sorting in spectrographs
- Prevention of focal plane stray light structure

Of particular concern for space instruments are the following areas:

- Contamination and cleaning issues
- Black surface selection and degradation of surfaces
- Radiative cooling issues and radiator design
- Bulk scatter in transmissive elements and grating scatter issues
- Specification of surface finish in optical elements
- Ghosting and filter issues
- Blackbody and calibration surfaces

Does the examination of the above issues ensure a system optimized against stray light concerns? The answer, unfortunately, is no. Many of the issues are interconnected and various tradeoffs would have to be conducted for system optimization. In many cases there is no time to identify the optimal design since the primary goal is to meet the system performance requirements in the most economical and conservative way possible. Often it is not possible to isolate these issues sufficiently well for individual trade studies. In this case, a preliminary design is arrive upon and then evaluated. The next section describes a way to verify the performance of the initial design.

4. Analysis of Stray Light

This paper has now outlined some general aspects of designing for stray light control. Although some space projects terminate their concern for stray light at this point, this is really only the first step. There are two more phases necessary to ensure success. The most important of these is the analysis of the system using a stray light analysis code. Many space instruments have had stray light analyses and have benefited from this process. For example, the designs of infrared optical systems such as IRAS, SIRTF, and ISO have been analyzed. Stray light analysis is not just for space systems; it is for any high performance optical system. Many airborne systems, laboratory spectrometers, illumination systems, and large scale physics experiments (such as laser interferometric detection of gravitational waves) have been analyzed using stray light analysis codes. Ground-based telescopes have also been analyzed to optimize their performance in the visible and thermal infrared (Pompea et al., 1992; Dinger, 1992).

Often a system "health check" or preliminary stray light analysis is done for the early or prototype design. A more detailed analysis can build upon this initial work later in the program when the design is finalized, but before metal is cut. In many cases, the analysis efforts have revealed areas where the system performance can be substantially improved. In some cases, catastrophic flaws in the design have been found.

Three major and several minor software packages are currently used for stray light analysis. The three major ones (in order of their creation dates) include GUERAP, APART, and ASAP. There is no current review of the software available, but the last review which was reasonably complete was by Breault (1990). The most flexible of these three codes in analyzing unusual or complex optical systems is probably the ASAP (and ASAP Plus) code, but APART is considered by many analysts to have the greatest power. A new interface for GUERAP has made it much easier to use. The most current version of GUERAP is called GUERAP V.

GUERAP V, from Lambda Research Corp. provides Monte Carlo raytrace simulation of flux propagation in imaging and non-imaging optical systems. It runs on a PC and is menu driven. ASAP Plus, from Breault Research Organization, runs on Sun, H-P, and RISC6000 workstations, VMS-based machines, and PCs. It is designed to be used with CAD files so that complex surfaces can be easily created. APART/PADE, also from Breault Research, uses a deterministic approach to analyze scattered, diffracted, and thermally emitted radiation. It runs on VAX/VMS platforms, H-P workstations, and PCs.

The non-deterministic codes such as GUERAP V and ASAP trace rays to verify the optical design, create important objects in the system, and send large numbers of rays through the system from a variety of angles. The level of detail needed in the modeling is an important topic by itself. Since only one entrance ray in 10^{13} may be incident at the focal plane, intelligent sampling (importance sampling) must be used to adequately represent the stray light paths.

In most codes, reflected, transmitted and scattered rays can be traced forwards and backwards through the system. The analysis shows the critical paths to the detector so that these can be removed to the extent possible. Ideally, after each major system modification, the analysis code is rerun to see if any new paths have been introduced and to verify that the changes have been beneficial.

A stray light analysis model provides a system level model where tradeoff performance studies of different designs and coatings can be made. The model can also aid in operations planning and decision making. In many systems, the observational limits on how close the sun, earth, and other off-axis sources can be to the telescope axis are determined by the stray light analysis.

One fundamental result of a stray light study is the Point Source Normalized Irradiance Transmittance (PSNIT) curve (Figure 2). The PSNIT curve defines the system stray light performance as a function of off-axis angle. It is a measure of the stray light rejection as a function of off-axis angles. A carefully constructed PSNIT curve provides a means of defining the performance and can help the analyst trace the paths that degrade system performance. The general shape of the curve gives the overall trend in the system performance with off-axis angle. Spikes in the curve represent areas that are significantly worse that the general stray light behavior of the system. These paths, represented by the letters A through G, can then be analyzed and hopefully blocked.

Another useful plot for array detectors is the distribution of stray light on the focal plane for a given off-axis angle (Figure 3). This plot is useful for predicting actual system performance and how stray light will affect background subtraction or other actual data characteristics. The plot can help the analyst determine whether the stray light power is distributed uniformly across the focal plane or concentrated in one or a few bright spots. However, since bright spots may occur at very specific angles, the computation of focal plane maps with sufficient angular resolution for a wide range of angles is difficult.

The emphasis is still on preventing bright spots in the focal plane through path analysis rather than by the construction of focal plane stray light maps at many off-axis angles. Figure 3 shows the scatter distribution across a detector for an off-axis source angle of 1.75 degrees. The scatter levels across the focal plane vary by a factor of ten in this particular case.

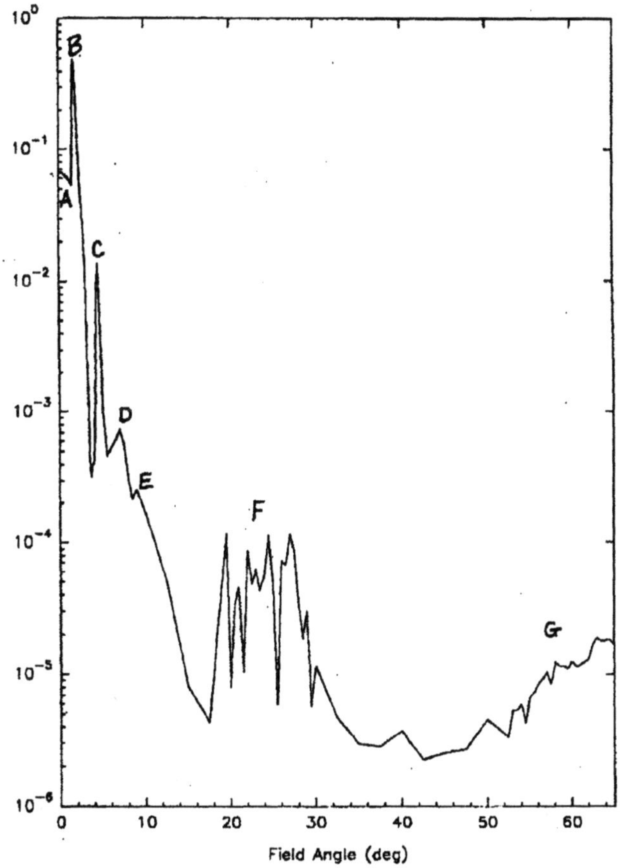

Fig. 2. The Point Source Normalized Irradiance Transmittance is plotted for various off-axis angles for a telescope system. The spikes are found at angles where stray radiation paths are important. These paths can then be blocked, or the objects that maintain these paths can be moved.

5. Testing for Stray Light

Why not simply design the best possible system, avoid the analysis step, and then test its performance? This build and test approach has a number of strengths, but some potentially severe weakness as well. In fact, the build and test approach is highly complementary in its strengths and weak points to the approach of stray light analysis by computer. For example, the real misalignments and real bidirectional reflectance distribution functions

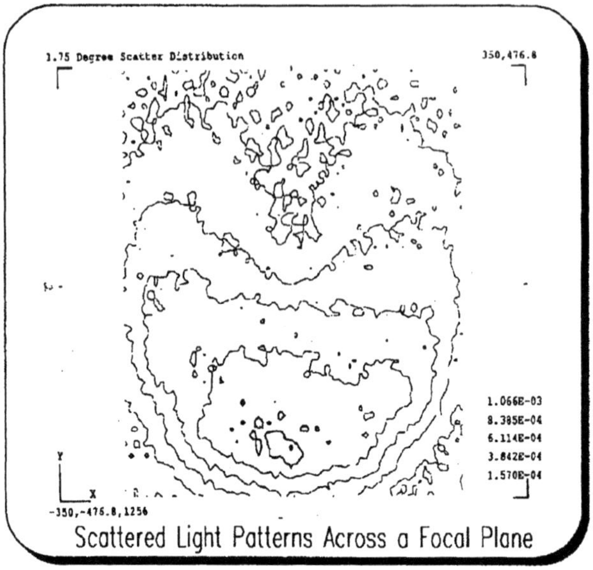

Fig. 3. If scattered light was a uniform, constant value across a focal plane array, it could be subtracted. However, most stray light paths gives a non-uniform distribution of excess irradiance that varies with the nature and angle of the source. This non-uniform distribution is difficult to subtract as it is difficult to fully characterize.

(BRDFs) of system surfaces are used in the "build and test" approach. The strengths and weakness of analysis and the build and test are summarized in table 1. Although the build and test approach seems quite appealing, a closer look at what it provides and what resources are needed to use it effectively make it quite expensive. For most high performance systems, testing is increasingly difficult, and testing late in the program is problematical. If the system does not pass, what can be done at this stage? The options are often limited.

Some rudimentary testing is certainly valuable. For example, with a camera at the focal plane, a few simple tests of off-axis performance can be performed with high intensity sources outside the system, if the design of the system will allow this. Some of the potential tests involve putting small dark or light objects in the object plane and moving them around to look for changes in the image plane. Much can be learned from this crude testing. This testing can also serve as a final check. However, this approach is probably best used as a preliminary check on a prototype system. This testing is most effective when there are serious flaws (from a stray light point of view) in the system design. Of course, a system "health check" done on the prototype design would probably have revealed these problems much earlier.

TABLE I

Comparison of "*build and test*" approach to "*system analysis*" approach (adapted from Breault, 1994).

Build and Test Approach Strengths	**Build and Test Approach Weaknesses**	**Analysis Approach Strengths**	**Analysis Approach Weaknesses**
Real BRDFs	Questions about accuracy of results & experimental setup	Answers are congruent with models	Programmer error a possibility
Real, assembled system performance	If first model fails, then what?	Redesign is easy in early stages	Modeling limitations
Misalignments included	What improvements are to be made?	Very informative	BRDFs of samples only
	Expense of creating model	Less expensive than hardware model	Expense of creating analytic model
	Expense of "marching army" at late program stage	Done early in program	

If there are specifications for stray light and overall system performance, then the object of the testing should be to measure the values of each. This is very difficult to do with high performance systems. Ideally, both extensive analysis and some testing would be done on every system. The consensus is that the analysis should be done early in the program to give the system the greatest chance of meeting its performance goals.

The stray light performance in space should be measured and compared with the results of analysis or ground tests. A critical comparison can yield rich insights on how best to use the system or in how to design the next system.

6. Black Surfaces and Contamination

A third key to preventing stray light problems is to treat black surfaces as important optical elements in your system. Benign neglect of such surfaces is a recipe for disaster. The selection of such surfaces is a subspecialty of the stray light area and is often a complicated process (Pompea and McCall, 1992; Pompea and Breault, 1994). "Black" surfaces play a number of roles in infrared and submillimeter systems. The primary role described here is for attenuation of off-axis radiation. Another important role is in calibration sources such as blackbodies. Although blackbody or radiator design is not really a stray light issue in most cases, the materials issues involved often send these choices to the stray light or spectrally selective surface expert.

Black surfaces are surfaces of low reflectivity that are used to attenuate stray light paths, usually as a coating on vane and baffle surfaces. The term "black" is used relative to the wavelength of interest and is not related to the appearance of the surface to the eye. The use of black, absorbing surfaces also creates emitting surfaces with potentially damaging consequences. Both the absorbing and emitting properties of surfaces and the paths of emitted radiation through the system must be examined in detail.

The selection of black surfaces and the prevention of contamination in space optical systems are often worked together, since black surfaces can be a major source of contamination. Particles can be shaken from surfaces by launch loads and outgasssing of black surfaces can contaminate other optical surfaces. Since black surfaces have a large surface area, even small percentage levels of particle shedding or outgassing from a given baffle material may have dire consequences for the performance of the optical system.

Black surfaces are also used in blackbody calibrations devices and in space radiators. For blackbody calibrators, a crucial issue is the stability of the emittance of the surface with time. In space radiators, the long-term performance of the radiator determines the radiator sizing, since the end-of-life performance must be maintained.

The difficulty of selecting a black surface for a blackbody, baffle, or radiator is often underestimated. This has led to substantial problems in a number of space instruments. Some of the potential pitfalls are discussed in Pompea and McCall (1992) and Pompea and Breault (1994). There are a large number of candidate surfaces to consider. Optical and environmental factors must be considered in detail. A few of the factors are measures of optical performance (e.g., reflectance, spectral emissivity, and scatter), outgassing, particle generation, and resistance of a surface to ultraviolet and atomic oxygen effects.

Finding the appropriate surface to use is not trivial. Well documented optical and environmental data is only available in the literature on a limited number of surfaces. Most of the data comes from a network of colleagues who collect data for their own private databases. Unfortunately, most bad experiences with a particular coating remain unpublished (McCall et al., 1992a). Although the choice of black surfaces cannot substantially improve the performance of a system poorly designed for stray light control, the wrong choice can jeopardize the whole system performance if the surfaces degrades under the space or launch environment.

When system performance expectations are high, the selection of surfaces becomes more complicated and more critical. In an optimized system, the choice of black surfaces can make a tremendous difference (Bergener et al., 1984; Pompea et al., 1989). With higher performance expectations for systems, there is a push to use more specialized surfaces in the system. Each area of the system may require a different type of surface (Pompea et al., 1989; Pompea and Breault, 1994).

For space systems, it is important to realistically model the baffle surfaces in the stray light analysis programs. The usual way to describe these surfaces is through the Bidirectional Reflectance (or Scatter) Distribution Function, called the BRDF (or BSDF) measured on a scatterometer (see Anderson et al., 1989 for a detailed description of one such scatterometer). The BRDF describes the scatter at a particular wavelength for a given input angle and scatter angle. To be effective, the BRDF should be measured at all wavelengths of interest. Because of the expense of making measurements, there is an effort to create an international database of measurements (McCall et al., 1992b). For realistic performance predictions, the effect of contamination or surface degradation must also be included. This means that accurate models of contaminated surfaces must be developed and compared against actual measurements.

At longer wavelengths, the surfaces are not as absorbing and the scatter properties of the surfaces are used to prevent specular propagation of radiation through the system. For example, surfaces such as Infrablack and Ames 24E rely on scattering as well as absorption to minimize propagation of radiation at wavelengths longer than 100 microns. Although many measurements have been made at 10.6 microns (Pompea et al., 1989), the scatter properties of black surfaces at far infrared and submillimeter wavelengths are not well documented. Beyond 10 microns, there is a dearth of BRDF measurements and many of the measurements are classified. Often, the specular reflectance as a function of wavelength is used as a first criterion for choosing surfaces. Very rough, thick surfaces are used to create moderate absorption and lambertian scattering of radiation (Pompea and Breault, 1994).

7. Management and Stray Light

It may seem presumptuous to state that stray light is first a management issue, and a technical issue only secondarily. However, stray light cannot be properly addressed as a technical issue without significant management support. This is particularly true since stray light is most effectively addressed early in a program, when there are many difficult technical issues and challenges that managers are addressing and prioritizing. Unless the value of stray light study and analysis are appreciated early, many systems will not be optimized from a stray light perspective even with extensive effort later in the program.

The long development time of space missions and the often sporadic funding allocated to them has often led to a linear development approach where requirements are defined this budgetary period, a preliminary optical design is done later (in another budgetary period), and then at a further stage the optical design is integrated into an opto-mechanical design. This leaves the systems analysis towards the end of or even after the design phase. A far more effective approach is simultaneous development by a small competent team of all of these aspects in parallel. This has the desired effect of examining key systems issues early into the instrument design. In this scenario, the stray light issues would be addressed very early in the program, in the initial sketches of the design. The preliminary optical design and the preliminary packaging approach would emerge at the same time.

In this scenario, the stray light expert (internal or external consultant) visits the program at key times, particularly early in the program, so that design changes can be accommodated. If there is a need for a development effort for the baffle surfaces or to address particular contamination issues, this can be acknowledged at this early stage.

This programmatic design for stray light control must start at the beginning of the program and relies upon having the key managers adopt a serious proactive approach to stray light issues. Many programs have tried a reactive approach and these have generally proven to be expensive. The stray light and contamination control plans are critical to mission success and must be treated early as critical paths from both a budgetary and planning perspective.

8. Summary

Stray light is a ubiquitous problem in infrared and submillimeter optical systems; the challenge is to control it to an acceptable level. The high performance space systems being planned will have many stray light problems not apparent in lower performance systems. One key to the prevention of stray light problems is early attention to the design from a stray light perspective. Another key to a successful program is to conduct a systems level analysis using stray light analysis software to identify problems and assess the magnitude of their effect on system performance. The third critical area is to do work early in the program on contamination control and the choice of surface coatings. The greatest current challenge in the stray light area is to apply what we know early enough in a program to be effective. This proactive approach is the most cost effective way to control stray light and optimize system performance.

References

Anderson, S., Pompea, S.M., and Shepard, D.F.: 1989, 'Performance of a Fully Automated Scatterometer for BRDF and BTDF Measurements at Visible and Infrared

Wavelengths' in *Proceedings of the SPIE: Stray Light and Contamination in Optical Systems*, **967**, 159.

Bergener, D. W., Pompea, S.M., Shepard, D.F. and Breault, R.P.: 1984, 'Stray Light Rejection Performance of SIRTF: A Comparison' in *Proceedings of the SPIE: Stray Radiation IV.*, **511**, 64.

Breault, R.P.: 1990, Stray Light Technology: Overview of the 1980 Decade (And a Peek into the Future)' in *Proceedings of the SPIE: Stray Radiation in Optical Systems*, **1331**, 2.

Breault, R.P. and Milsom, D.: 1992, 'Stray Light Analysis of the Cassini Telescope' in *Proceedings of the SPIE: Stray Light IV*, **1753**, 210.

Breault, R.P.: 1994, *Stray Light Course*, Breault Research Organization, Tucson, Arizona, unpublished.

Dinger, A.: 1992, 'Thermal Emissivity Analysis of a Gemini 8-Meter Telescope Design' in *Proceedings of the SPIE: Stray Light IV*, **1753**, 183.

McCall, S.H.C.P., Pompea, S.M., Breault, R.P. and Regens, N.L.: 1992a, 'Reviews of Black Surfaces' in *Proceedings of the SPIE: Stray Light IV*, **1753**, 158.

McCall, S.H.C.P., Sinclair, R.L., Pompea, S.M. and Breault, R.P.: 1992b, 'Spectrally Selective Surfaces for Ground and Space-Based Instrumentation: Support for a Resource Base' in *Proceedings of the SPIE: Space Astronomical Telescopes and Instruments II*, **1945**.

Pompea, S.M., Shepard, D.F. and Anderson, S.: 1989, 'BRDF Measurements at 6328 Angstroms and 10.6 Micrometers of Optical Black Surfaces for Space Telescopes' in *Proceedings of the SPIE: Stray Light and Contamination in Optical Systems*, **967**, 236.

Pompea, S.M., Mentzell, J.E. and Siegmund, W.E.: 1992, 'A Stray Light Analysis of the Sloan Digital Sky Survey Telescope' in *Proceedings of the SPIE: Stray Light IV*, **1753**, 172.

Pompea, S.M., and McCall, S.H.C.P.: 1992, 'Outline of Selection Processes for Black Baffle Surfaces in Optical Systems' in *Proceedings of the SPIE: Stray Light IV*, **1753**, 192.

Pompea, S.M. and Breault, R.P.: 1994, 'Optical Black Surfaces' in *Handbook of Optics*, 2nd edition, Optical Society of America.

PULSE TUBE REFRIGERATORS

A. RAVEX, L. DUBAND AND P. ROLLAND
CEA / Département de Recherche Fondamentale sur la Matière Condensée
Service des Basses Températures
17 rue des Martyrs
38054 Grenoble Cedex 9 - FRANCE

Abstract.
Interest in the pulse tube comes from its potential for high reliability and low level of induced vibration.
A numerical model has been developed to provide a tool for practical design. It has been successfully validated against the experimental results obtained with a single stage double inlet pulse tube which has achieved a temperature of 28 K at a frequency of a few Hz.
Further developments have demonstrated the capability of operating a pulse tube at higher frequencies in association with a Stirling pressure oscillator.
Current projects include coaxial geometry for miniature pulse tubes with linear resonant pressure oscillators. A 4 K multistaged pulse tube is also in development.

1. Introduction: Pulse tube refrigerators description

The basic pulse tube refrigerator was first described in 1964 by Gifford and Longsworth. It is shown schematically in Figure 1.

The pulse tube itself is a thin walled cylinder with heat exchangers located at each end. It is supplied through a regenerator with pressure waves produced either by a pressure oscillator or by a three way distributor associated to a compressor. The cooling effect relies on heat exchange between the gas and the tube wall known as surface heat pumping (Wheatly et al., 1985; Richardson, 1986). Such a device operates at low frequencies (1–5 Hz) and has achieved a low temperature of 124 K (Longsworth, 1967). The low efficiency of this basic design was certainly the main reason why the pulse tube remained for a long time undeveloped.

In 1984, Mikulin et al. modified the basic pulse tube design by adding a valve (V1) and a reservoir volume at its closed end. They reached a temperature of 105 K using air as the working fluid and soon afterward Radebaugh et al. (1986) reached 60 K using helium. This new design shown schematically in Figure 1 is referred to as the orifice pulse tube.

An analytical model describing the behaviour of the orifice pulse tube has been developed by Storch & Radebaugh (1988) to gain a better understanding of the refrigeration process.

This enthalpy flow analysis is based on the first law of thermodynamics: an energy balance on a control volume at the cold end of the pulse tube shows that the time averaged enthalpy flow over a period τ, assuming an

Space Science Reviews **74**: 195–203, 1995.

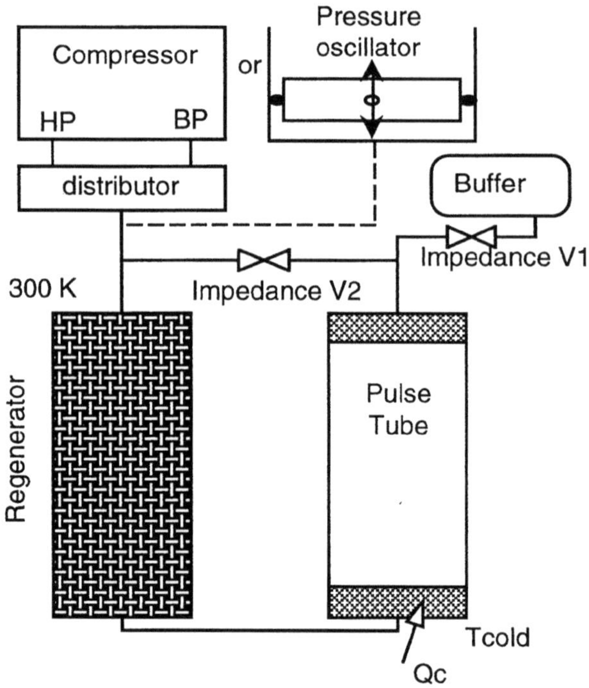

Fig. 1. Schematic of a pulse tube refrigerator.

ideal gas behaviour, is given by:

$$\langle \dot{H} \rangle = \frac{C_{\rm P}}{\tau} \int_0^\tau \dot{m} T \mathrm{d}t$$

where $\dot{m}$ is the mass flow rate, $C_{\rm p}$ is the specific heat of gas at constant pressure and T the gas temperature.

Since the mass flow rate m is equal to $\rho A_{\rm pt} u$ where ρ is the density of the gas equals to $\rho = P/rT$ for an ideal gas (P: pressure), u is the local gas velocity and $A_{\rm pt}$ is the cross section area of the pulse tube, the time-average enthalpy flow rate can be written as:

$$\langle \dot{H} \rangle = \frac{C_{\rm P} A_{\rm pt}}{r\,\tau} \int_0^\tau u P \mathrm{d}t$$

The net cooling power $\dot{Q}_{\rm c}$ is then given by the following expression:

$$\langle \dot{Q}_{\rm c} \rangle = \langle \dot{H} \rangle - \langle \dot{H}_{\rm r} \rangle$$

where $\langle \dot{H}_{\rm r} \rangle$ is the average enthalpy flow from the regenerator.

If the cyclic variations of pressure and velocity are assumed to be sinusoidal, the phase shift angle ϕ between them is the important parameter governing the cooling process. A direct comparison with Stirling cycle can thus be done. In a Stirling cryocooler the displacer/expander in the cold finger leads the motion of the pressure oscillator piston by about 90°. As a result the mass flow rate (i.e. gas velocity) and the pressure at the cold end are approximately in phase leading to the expected refrigeration effect. In a pulse tube the proper phase relationship is obtained by the adjustment of the orifice impedances and reservoir volume at room temperature.

Although the orifice pulse tube refrigerator has a greatly improved efficiency in comparison with the basic pulse tube, the mass flow rate through the regenerator is still very large in comparison with a Stirling cryocooler and consequently the specific cooling power and ultimate temperature achieved aren't as good. This is mainly due to the fact that a large volume of gas with no refrigerative effect flows through the regenerator into the pulse tube because of the pressure oscillations. Zhu et al. (1990*a*) have suggested the double inlet pulse tube concept to overcome this disadvantage and they reached a temperature of 42 K (Zhu et al., 1990*b*). As schematically shown in Figure 1 the second valve (V2) directly connects the hot end of the pulse tube to the pressure wave generator in this configuration.

Due to these successive improvements in performance the interest in the pulse tube refrigerators has grown rapidly in the last few years. Because they have no moving components in the low temperature region, they have the potential for high reliability and low vibration at the cold tip which are of major importance in satellite applications.

Recently Tward et al. (1990) have reported a system efficiency for a double inlet pulse tube operated with a flexure bearing Oxford type pressure oscillator which is very close to that of a comparable Stirling cryocooler.

2. Theoretical modelisation and experimental verification

We have undertaken an experimental characterisation and a thermal modelisation of the double inlet refrigerator to get a better understanding of its operation and a practical tool for further system sizing and efficiency calculations (Liang, 1993).

2.1. Experimental set up and results

The pressure oscillation is generated by an helium compressor connected to the pulse tube by way of an electromagnetic 3 way solenoid valve.

The regenerator consists of a stainless steel tube (18 mm inner diameter, 170 mm long) filled with 180 mesh stainless steel wire gauze discs.

Several pulse tubes (stainless steel tube 200 mm long) with inner diameter ranging from 10 mm up to 20 mm have been tested. At both end of the pulse tubes copper gauze discs are brazed for flow straightening and heat exchange.

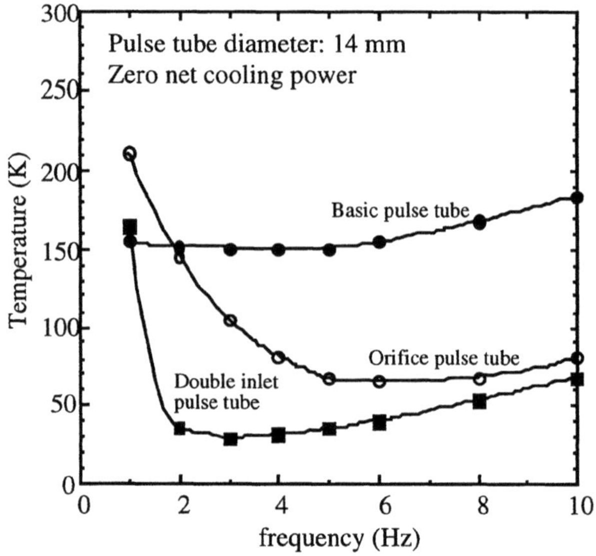

Fig. 2. Ultimate temperature versus frequency

At room temperature a buffer volume is connected to the pulse tube through an adjustable needle valve (V1) and a by pass adjustable needle valve (V2) is also inserted between the warm ends of the regenerator and the pulse tube. These two valves allow for tests in any type of configuration (basic, orifice or double inlet pulse tube).

The ultimate cold end temperature (no net cooling power) has been measured for various pulse tube diameters, pressure wave frequencies and opening of the needle valves. Typical experimental results corresponding to the optimised needle valves adjustments are reported in Figure 2. An ultimate temperature $T = 28\,\mathrm{K}$ has been obtained in the double inlet configuration for the 14 mm inner diameter pulse tube at a frequency of 3 Hz with a pressure wave amplitude $\Delta P = 7\,\mathrm{bar}$ and a mean pressure $P = 12\,\mathrm{bar}$.

Cooling power measurements are also reported in Figure 3. They have been performed at frequencies and valve opening adjustments corresponding to the lowest temperature previously achieved with no net cooling power. In the double inlet configuration a net cooling power of 16 W at 80 K has been obtained.

The influence on the cooling performances of several parameters such as geometry of the pulse tube, regenerator mesh gauze disc matrices, pressure oscillation amplitude and frequency, mean pressure and valves opening adjustment has been systematically studied for further comparison with theoretical calculations from the theoretical model.

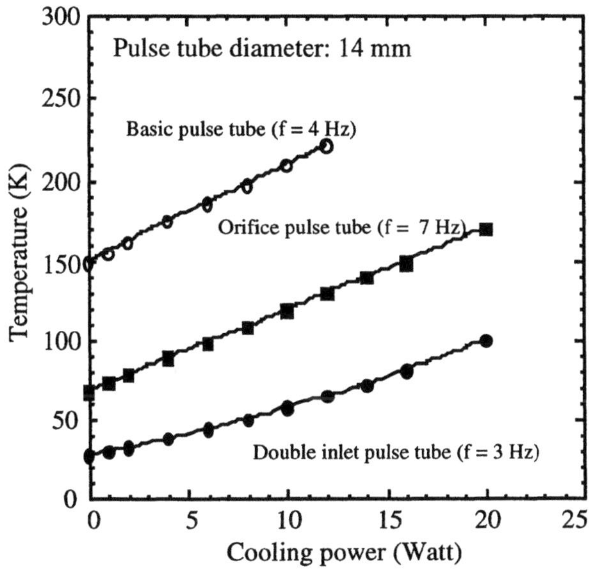

Fig. 3. Cooling power for various configurations

2.2. Modelisation

We have developed a numerical model to describe the double inlet pulse tube refrigerator based on the conservation of mass and first law of thermodynamics. The helium is assumed to behave as a perfect gas.

The gross refrigeration is calculated taking into account two independent contributions: the enthalpy flow in the bulk of the gas as described by Storch & Radebaugh (1988) and the enthalpy flow due to heat exchange with the wall as described by Wheatley (1985) or Richardson (1986). The thickness of the layer in which the surface heat pumping effect occurs is calculated taking into account the thermal diffusivity of helium and solving radial transient heat transfer equations. In the remaining central volume an adiabatic plug flow is assumed.

Compression in the compressor or pressure oscillator is assumed to be isothermal. The hot and cold heat exchangers in the pulse tube are assumed to be perfect. The needle valves characteristics have been experimentally determined and the appropriate pressure drop relation versus mass flow rate is introduced in the calculation.

To determine the net cooling power from the gross refrigeration, parasitic contributions should be determined and subtracted. The thermal conduction through the pulse tube and regenerator stainless steel walls as well as through the regenerator metallic screens (empirical law) are thus calculated and taken into account. The contribution to the thermal loss resulting from the regenerator thermal inefficiency is calculated with a special subroutine developed in our laboratory in which the theoretical correlations for heat transfer in metallic screen packages proposed by Kays & London (1984) are used. Related correlations for pressure drop are also used to determine the pressure oscillation amplitude attenuation through the regenerator resulting in a gross cooling power reduction.

Results predicted by this model are found to be in fairly good agreement with experimental data. For example, in Figure 4 the measured net cooling powers at 100 K for various valves opening and as a function of frequency are compared with calculated data.

This model validated by the experiment will be an efficient tool for further design and optimisation of pulse tube refrigerators.

3. Miniature pulse tube coolers

Miniature Stirling coolers are presently widely used for infrared detectors cooling. The strong requirement for high reliability has lead to many technological improvements: linear motor drives, clearance seals, frictionless bearings. But Stirling cryocoolers still have two moving parts: the pressure oscillator piston and the cold finger displacer. In the pulse tube refrigerator the moving displacer has been eliminated conferring a potential for lower cost, higher reliability and less vibration.

In a preliminary attempt to validate these assumptions we have developed a miniature pulse tube associated with a pressure oscillator adapted from a commercial oil free piston air compressor.

Some characteristics and performances of this prototype operating in an orifice pulse tube configuration are summarised in the following table.

Oscillator swept volume	: 8 cm^3
Mean pressure	: 1.5 MPa
Frequency of operation	: 25 Hz
Ultimate temperature	: 80 K
Typical cooling power	: 1 W @ 110 K

Mean pressure and pressure oscillations were limited by the rotating motor torque and the by pass flow in the piston/cylinder clearance, but high frequency operation was demonstrated.

Later on, in the framework of a collaboration with CRYOTECHNOLOGIES S.A., different miniature double inlet pulse tubes have been designed and tested with a standard Stirling cooler pressure oscillator.

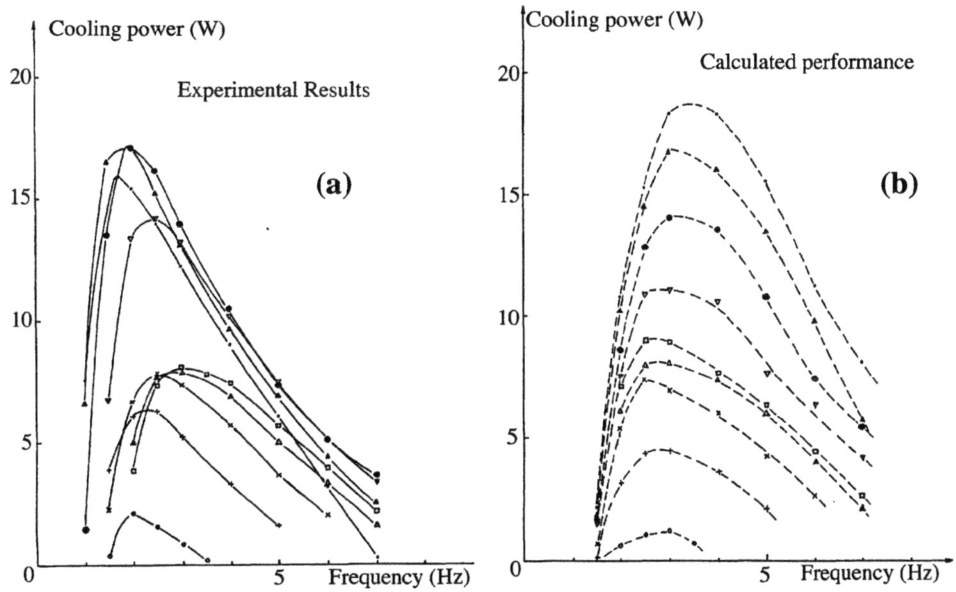

Fig. 4. Comparison between experimental and calculated performance. The various curves correspond to different valves opening.

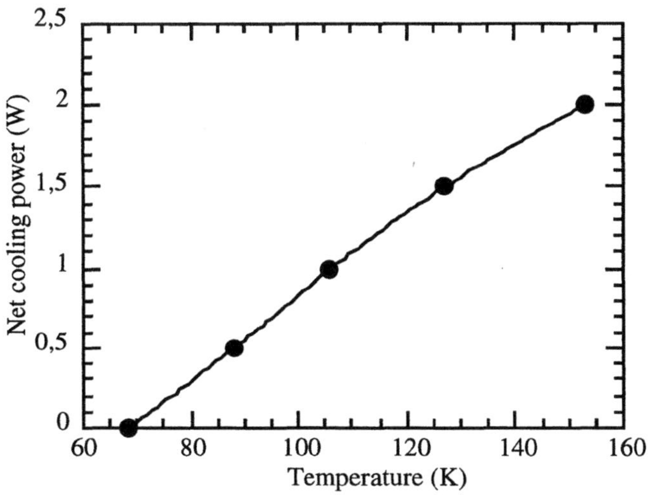

Fig. 5. Experimental cooling power of a miniature pulse tube prototype. The full lines are guide for the eyes.

Figure 5 shows as an example the net cooling power versus the cold tip temperature for one of these prototypes. The good agreement between experimental and theoretical data validates our previously described model in a very different frequency and cooling power range. The main characteristics of the pulse tube configuration corresponding to these results are summarised here after.

Pressure oscillator swept volume	: 2.1 cm^3
Mean pressure	: 3.5 MPa
Operation frequency	: 20 Hz
Ultimate temperature	: 67 K

Further work is being done with this pressure oscillator to estimate the performances of a coaxial geometry for the pulse tube and regenerator. Efficiency measurements are also planned recording PV diagram with a modified oscillator.

4. Future work

The preliminary results presently available and reported here gives us a great confidence in the potentialities of pulse tube refrigeration. New developments are currently underway in our laboratory.

The further step in the improvement of the "Stirling like" pulse tube refrigerator is the use of a linear pressure oscillator. The necessary resonant operation will induce a constraint on the sizing and optimisation of the system which will be included in our model. This work is presently underway and a double inlet pulse tube will be soon associated with a commercial linear Stirling pressure oscillator. The use of highly efficient compressor developed for space borne applications is also to be considered.

"Gifford Mac Mahon like" (i.e. using helium compressor associated with a distributor) pulse tube are also under development. In the low frequency range the goal is to achieve large cooling power at intermediate temperature or low ultimate temperature with multistaged systems. Recently in a preliminary test, a multistaged pulse tube using magnetic material for regeneration has reached a 5 K temperature. Concurrently a theoretical work has been initiated to modelise the thermal behaviour and efficiency of the regenerator in a temperature range [20 K – 4 K] where specific heat of the regenerative material exhibits a sharp anomaly and the helium no more behaves as a perfect gas.

5. Conclusion

A development program on pulse tube refrigeration has been undertaken at CEA/SBT. A numerical model has been developed and successfully validated against experimental data obtained with large capacity/low frequency (16 W/80 K at 3 Hz) and miniature high frequency (1 W/100 K at 25 Hz) prototypes. This numerical tool now available for design and optimisation is currently being used for new developments. These developments include coaxial geometry pulse tube associated with resonant linear pressure oscillators, or low temperature multistaged systems.

The association of a pulse tube with a pressure oscillator developed for long life space cryocooler will certainly contribute to simplify and improve the integration and reliability aspects.

References

Gifford, W.E., Longsworth, R.C.: 1964, *Trans. ASME J. Eng. Ind.* **63**, 264.
Kays, W.M., London, A.L.: 1984, '*Compact Heat Exchangers*', 3rd edition, New York, Mc Graw Hill Series in Mechanical Engineering,
Liang, J.: 1993, '*Development and Experimental Verification of a Theoretical Model for Pulse Tube Refrigeration*', Thesis, Grenoble.
Longsworth, R.C.: 1967, *Adv. Cryo. Eng.* **12**, 608.
Mikulin, E.I., Tarasov, A.A., Shkrebyonock M.P.: 1984, *Adv. Cryo. Eng.* **29**, 629.
Radebaugh R., Zimmermann, J., Smith, D.R., Louie, B.: 1986, *Adv. Cryo. Eng.* **31**, 779.
Richardson, R.N.: 1986, *Cryogenics* **26**, 331.
Storch, P.J., Radebaugh, R.: 1988, *Adv. Cryo. Eng.* **33**, 851.
Tward, E., Chan, C.K., Burt, W.W.: 1990, *Adv. Cryo. Eng.* **35**, 1207.
Wheatly, J., Hoffer, J., Swift, G.W., Migliori, A.: 1985, *Am. J. Phys.* **53**, 147.
Zhu, S. W., Wu, P. Y., Chen, Z. Q.: 1990*a*, *Cryogenics* **30**, 514.
Zhu, S.W., Wu, P. Y., Chen, Z.Q.: 1990*b*, *Cryogenics* **30**, 257.

THE USE OF CLOSED CYCLE COOLERS ON SPACE BASED OBSERVATORIES

T. W. BRADSHAW and A. H. ORLOWSKA
Rutherford Appleton Laboratory,
Chilton, Didcot, Oxon, OX11 OQX, UK

Abstract. Many proposed space based observations will rely on the use of closed cycle and passive cooling systems to provide the thermal environment for high sensitivity. The use of closed cycle mechanical coolers on space telescopes poses particular integration problems; some of these difficulties are discussed in this paper.

One of the major problems envisaged is that of exported vibration. This problem, and that of the heat sinking required, can be alleviated by siting the compressors of the Stirling cycle precooler further from the displacer unit. The effect of the separation between the compressors and the displacer on the performance of the Stirling cycle precooler has been measured. Increasing the separation from 170 mm to 565 mm decreases the cooling power at 25 K from 220 mW to 180 mW. In most applications this would be acceptable.

The pre-cooler provides cooling at a single point. In situations where refrigeration of extended objects (e.g. telescope mirrors) is required, some distribution method has to be found. A scheme for achieving this is presented together with preliminary calculations on such a system.

Temperatures in the region of 2.5 to 4 K are required to meet the requirements for long wavelength detectors. We have demonstrated how these temperatures can be achieved in a continuously operating closed cycle cooler that has been engineered for space applications. This cooler consists of a two-stage Stirling cycle precooling a closed cycle Joule-Thomson (JT) stage. Temperatures in the region of 4 K are achieved by the use of helium-4 in the JT system. The lighter isotope of helium is used to obtain temperatures down to 2.5 K. Under no-load conditions the precooler reaches a base temperature of 11.3 K. The JT system achieves 4.3 K with a 10 mW heat load and 2.5 K with a heat load of over 3 mW. The input power to the cooler is approximately 126 W.

The temperature stability of the cooler at low temperatures is important to keep detector drift to a minimum. The temperature of the JT stage has been measured in uncontrolled laboratory conditions and found to vary by only 30 mK over a seventy hour period. The pre-cooler temperature varied by approximately 0.6 K during these measurements.

1. Introduction

Astronomy in the far infra-red poses significant problems for the instrument designer. The instrument and detectors have to be cooled to low temperatures to reduce the thermal background. In the past this has been achieved by using stored cryogens that are inherently life limited. A solution to this problem is to use closed cycle mechanical coolers. The coolers developed at the Rutherford Appleton Laboratory have been built using proven space technology and a number of the single stage 80 K coolers have been flown in space (Werret et al., 1986). The development of these coolers has continued and two-stage, 4 K and 2.5 K coolers are now available. The integration of these coolers on to an instrument is not trivial and a number of factors need to be taken into account from the outset:

Space Science Reviews **74**: 205–213, 1995.

- Exported vibration from the coolers
- The removal of the heat of compression
- The temperature and temperature stability of the detectors
- The distribution of the cold from the cold stage of the closed cycle cooler

The position of the compressors is crucial in order to minimise problems from the first two of these factors and it would be helpful to site the compressors as far away from the displacer as possible. This might also allow the compressors to be positioned on a cold plate for heat removal. Some measurements have been made on the effect of increasing the distance between the compressors and displacer.

The temperature stability of the 2.5 K cooler was measured in the laboratory with no environmental control. The degree of thermal control available on a spacecraft varies widely and depends on the capability of the bus, orbit and many other design factors. Some measurements are presented here which will give an idea of the likely magnitude of the temperature stability of the cold end so that the appropriate measures (if any) can be taken in the design of an instrument.

The removal of heat from baffles and optics is difficult to achieve with closed cycle coolers as they provide only spot cooling. Passive methods involving the use of busbars and heat pipes invoke mass and integration penalties. In some instances cooling is only required during certain phases of the project e.g. during initial cool-down. The EDISON observatory, for example requires cooling of the optics and baffles during the initial cool-down period to reduce the time to reach thermal equilibrium and to provide thermal stability thereafter. Most of the cooling is achieved through passive means. An active method of cooling, involving the use of a gas cooling loop, obviates many of these problems. In addition it also provides mechanical isolation from the item being cooled.

2. The two Stage Cooler

2.1. Mechanical arrangement

The two stage cooler consists of two compressor units with reciprocating pistons mounted so that the pistons move in opposition. These are connected to a displacer unit by a copper pipe. The two stage cooler provides cooling to temperatures below 20 K. Typical uses for this cooler on a telescope might be for cooling optics, radiation shields or baffles. It is also used as a pre-cooler for the 2.5/4 K cooler.

The reciprocating elements can be balanced in order to reduce the exported vibration to quite low levels (Ross et al., 1991), however, it would be prudent to site the compressors as far away from the detectors as possible. In most cases the benefits of easier heat removal and simpler integration would outweigh a small performance penalty in increasing the separation between the compressors and the displacer.

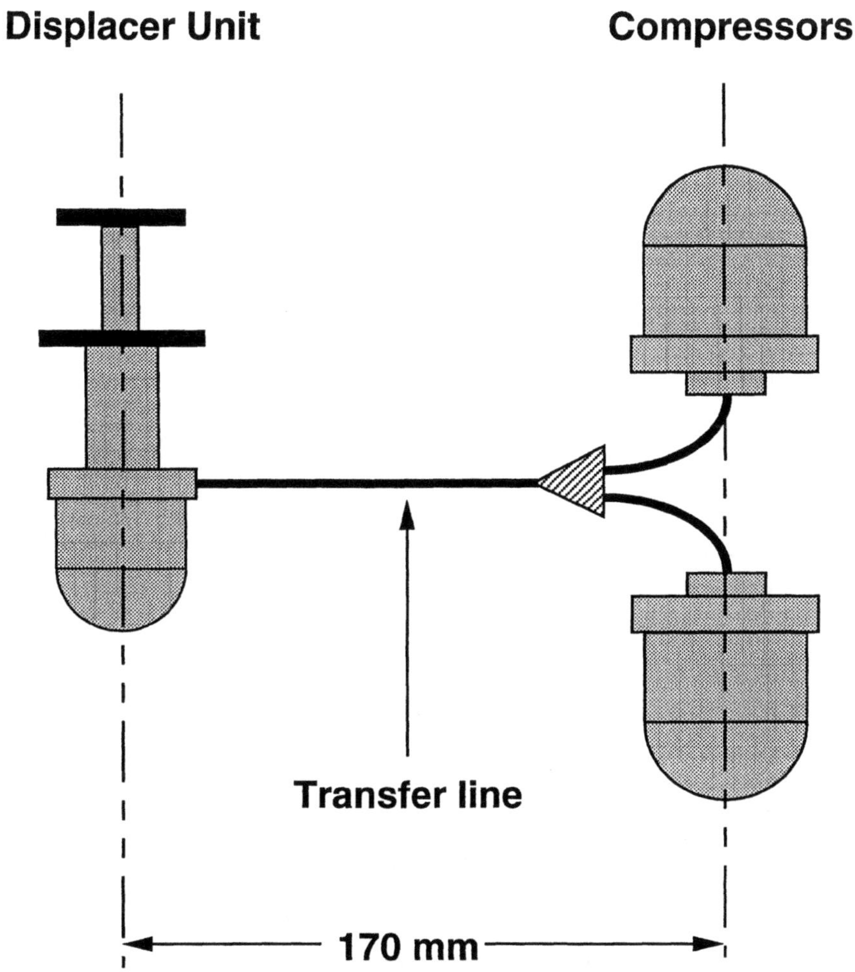

Fig. 1. The standard two-stage cooler.

2.2. Effect of increasing the transfer line length

In the original design the compressor-displacer separation was made as short as possible as shown in Figure 1. This configuration keeps the dead volume in the cooler to a minimum. Extension pieces were manufactured to increase the 170 mm separation by 185, 287 and 395 mm.

The cooler parameters (displacer stroke and phase) were optimised with a 100 mW heat load on the cold finger and the cooler fill pressure was 9 bar. The load lines for the various separation distances are shown in Figure 3.

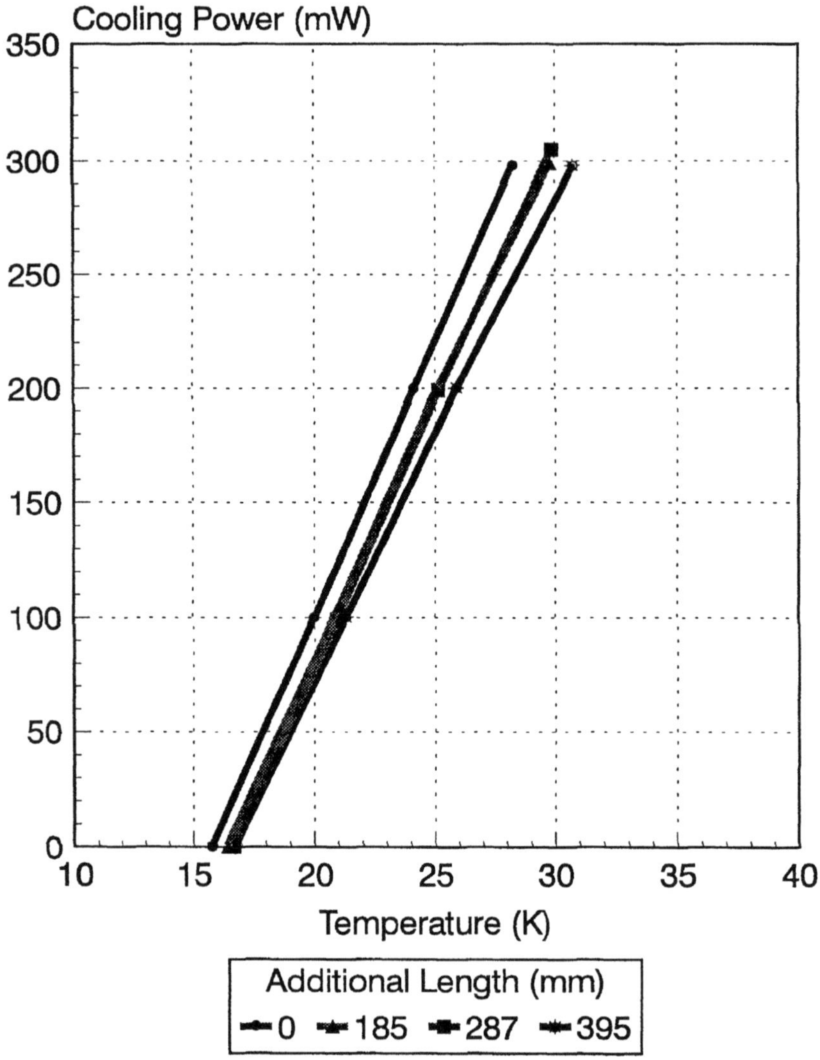

Fig. 2. The load lines at 9 Bar.

Other measurements indicated that the thermal efficiency of the cooler was not unduly affected by the extra dead volume. It appears that the compressors can be sited some distance from the displacer unit without a large performance penalty.

2.3. A SCHEME FOR DISTRIBUTION OF COOLING

The Stirling cycle cooler provides refrigeration at a single point, not the optimum arrangement for a large object. Using the technology developed for the closed cycle JT cooler, a method for cooling large objects by means of a gas cooling loop has been devised. A small pump, at room temperature, circulates helium gas around a cooling loop. Helium travels from the pump, through a counter current heat exchanger, to the cold stage of the Stirling cycle. It then flows through the loop which is thermally anchored to the cold stage of the Stirling cycle pre-cooler and the object to be cooled, returning through the heat exchanger to the pump. The pump could be similar to the JT compressors used in the 4 K cooler and could be at some distance from the cold part. The cooling loop itself could be any length, up to say 4 m, providing it was in a cold environment, shielded from thermal radiation.

The performance of a typical cooling loop is shown in Figure 3. This figure shows the cooling power of the loop and the minimum temperature (at the beginning of the loop) and the maximum temperature (where the gas returns to the cold stage of the cooler). The heat lift depends on the flow through the cooling loop, and the temperature. With higher gas flows larger heat lifts are possible, but at higher temperatures. It can be seen from a comparison of Figures 2 and 3 that there is a reduction in available cooling power due to heat exchanger ineffectiveness. Further design work could reduce this shortfall. The discontinuity at approximately 18 K is due to a laminar/turbulent transition in the heat exchanger calculations.

3. Cooling power Measurements

The 2.5 K cooler is a development of the 4 K cooler developed under ESA contract and with RAL funding (Bradshaw & Orlowska, 1988). This cooler was modified to run with helium-3 instead of helium-4 using a smaller JT orifice to reduce the flow to 1 mg/s. The temperature achieved in a JT system is determined by the pressure on the downstream side of the JT orifice. Improvements were made to the JT compressors to enable lower pressures to be reached.

The method used to measure the cooling power was as follows:

- A heat load was applied to the 2.5 K stage until all the liquid was boiled off. This is marked by a rapid increase in temperature.
- The heat load is then reduced to a low value and the temperature monitored closely. If the cooling power is greater than the heat load then the temperature will return to the base temperature.
- The process is repeated returning the heat load to a new value.

This process can be seen in Figure 4. The cooler returned to base temperature with heat loads of 0, 2.5, 3 and 3.5 mW. With a heat load of 3.5 mW

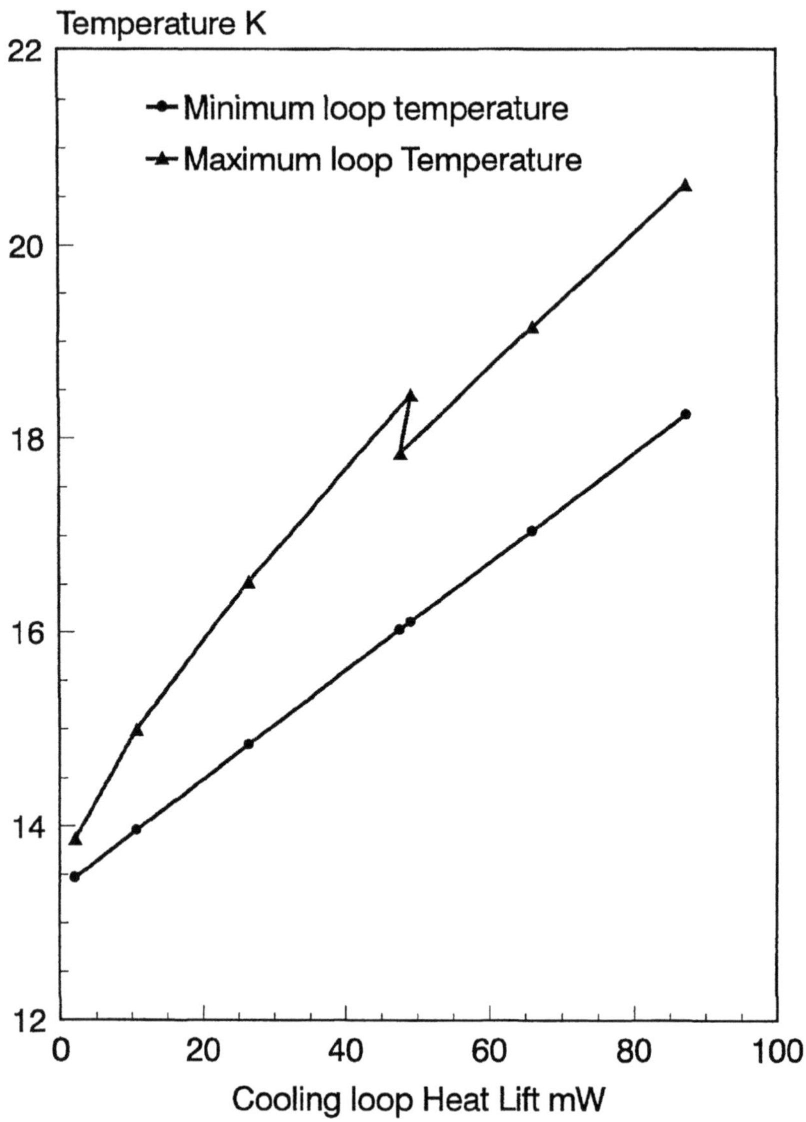

Fig. 3. Performance of the cooling loop.

the cooler temperature reached its base temperature very slowly, indicating that this was close to the maximum heat lift of the system.

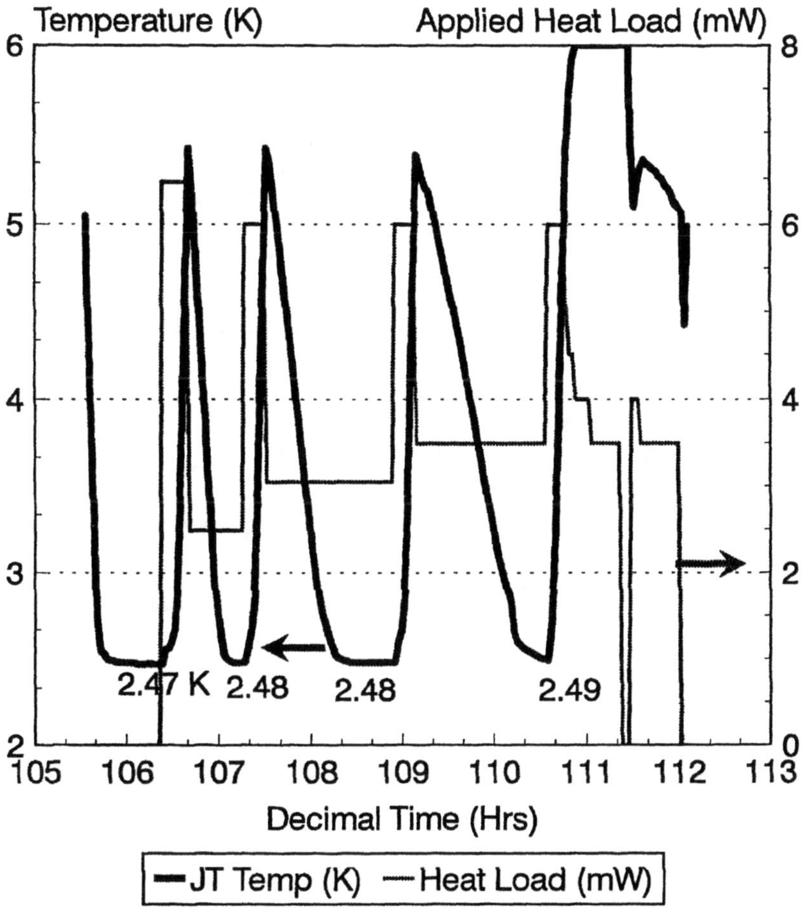

Fig. 4. Cooling power measurements.

4. Temperature stability of the 2.5 K Cooler

The cooler was left close to its base temperature for a period of seventy hours. The temperature of the pre-cooler and the 2.5 K plate was monitored over this period. The results are shown in Figure 5.

This figure shows the 2.5 K plate following closely the variation in temperature of the pre-cooler. No attempt was made to stabilise the temperature of the pre-cooler. The day-time temperature varied between 18 and 22 degrees C. The night time temperature was not monitored. The mass of the 2.5 K plate was of the order of 25 g.

Even rudimentary thermal control of the pre-cooler should reduce this variation to an even lower level.

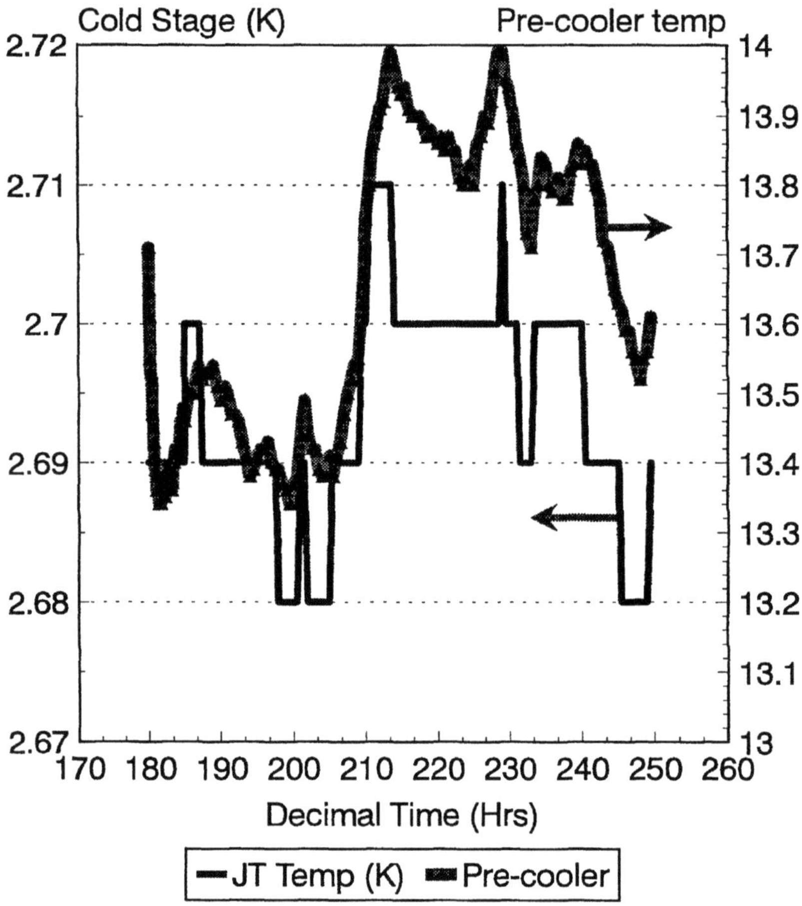

Fig. 5. The temperature stability of the cooler.

5. Summary and Conclusions

The use of closed cycle mechanical coolers on spacecraft is feasible as long as the implications of their use is understood and the instrument is designed with these in mind. Some methods of alleviating these problems have been presented.

The effect of increasing the transfer pipe length in order to physically separate the major sources of vibration (the compressors) from the cooled object was considered. It was shown that the heat lift of the two-stage Stirling cycle pre-cooler was degraded from 220 mW to 180 mW at 25 K. For most applications this degradation is acceptable. The addition of the extra pipe length did not introduce significant extra losses. The application of extra input power by increasing the compressor stroke (and/or pressure) will regain some of this performance, subject to the mechanical limits of the compressors not being exceeded. The positioning of the JT compressors is not critical since the JT system requires a "DC" pressure produced by rectifying the alternating pressure wave from the reciprocating compressors with reed valves. This pressure is not a function of dead volumes in the system outside the compressors, hence long transfer lines carrying the JT gas to and from the compressors can be accommodated.

The use of a gas cooling loop in order to refrigerate extended low thermal conductance objects was investigated and found to be a practical solution to the problems of spot cooling and would also provide increased mechanical isolation.

The temperature stability of the 2.5 K cooler was approximately 30 mK over a seventy hour period. Further investigations are required to map out the stability of the cooler with respect to changes in the ambient operating conditions.

The problems associated with closed cycle coolers are not insurmountable but further work is required in the areas of vibration reduction and compensation, thermal stability of the coolers and the implications of electromagnetic interference on bolometers and detectors.

Acknowledgements

The authors would like to acknowledge the support of ESA and the Science and Engineering Research council and the help of many individuals particularly Roger Wolfenden, Jamie Hieatt and Bill Blakesley for their support in this work.

References

Bradshaw, T. W., Orlowska, A. H.: 1988, 'A 4 K mechanical refrigerator for space applications' in *Proceedings of the third European Symposium on space thermal control and life support systems*, ESA SP-288/ISSN 0379/6556

Ross, R. G., Johnson, D. L., Kotsubo, V.: 1991, 'Vibration characterization and control of miniature stirling-cycle cryocoolers for space applications', *Adv. Cryo. Eng.* **Vol 37B**, 1019

Werret, S. T. et al.: 1986, 'Development of a small stirling cycle cooler for spaceflight applications', *Adv. Cryo. Eng.* **Vol. 31**, 791-799

ADIABATIC DEMAGNETISATION REFRIGERATORS FOR FUTURE SUB-MILLIMETRE SPACE MISSIONS

I. D. HEPBURN, I. DAVENPORT and A. SMITH
Mullard Space Science Laboratory,
Department of Space and Climate Physics,
University College London,
Holmbury St. Mary, Dorking,
Surrey, England

Abstract.
Space worthy refrigeration capable of providing a 100 mK and below heat load sink for bolometric detectors will be required for the next generation of sub-millimetre space missions. Adiabatic demagnetisation refrigeration (ADR), being a gravity independent laboratory method for obtaining such temperatures, is a favourable technique for utilisation in space.
We show that by considering a 3 salt pill refrigerator rather than the classic single salt pill design the space prohibitive laboratory ADR properties of high magnetic field (6 Tesla) and a < 2 K environment (provided by a bath of liquid ^{4}He) can be alleviated, while maintaining a sufficient low temperature hold time and short recycle time. The additional salt pills, composed of Gadoliniun Gallium Garnet (GGG) provide intermediate cooling stages, enabling operation from a 4 K environment provided by a single 4 K mechanical cooler, thereby providing consumable free operation. Such ADRs could operate with fields as low as 1 Tesla allowing the use of high temperature, mechanically cooled superconducting magnets and so effectively remove the risk of quenching.
We discuss the possibility of increasing the hold time from 3 hours, for the model presented, to between 40 and 80 hours, plus reducing the number of salt pills to two, through the use of a more efficient Garnet.
We believe the technical advances necessitated by the envisaged ADRs are minimal and conclude that such ADRs offer a long orbital life time, consumable free, high efficiency means of milli-Kelvin cooling, requiring relatively little laboratory development.

Key words: space – ADR – magnetic cooling – paramagnetic materials

1. Introduction

Current ground-based bolometric detectors used in sub-millimetre wave astronomy require cooling to 0.3 K (invariably through pumped ^{3}He cryostats) with the next generation of bolometric detectors begin cooled to 0.1 K (through either compact ^{3}He-^{4}He dilution refrigerators (Cunningham & Gear, 1990) or adiabatic demagnetisation refrigerators (ADR, Lesyna et al.)). If these detectors are to be used in space observatories then cooling to 0.1 K will be required.

Cooling to such temperatures has been possible since the early 1930s through the gravity independent process of adiabatic demagnetisation refrigeration. Kelvin and milli-Kelvin temperatures are obtained by the interaction of a magnetic field, now produced by superconducting magnets, and the magnetic moments of a paramagnetic material at liquid helium temper-

Space Science Reviews **74**: 215–223, 1995.

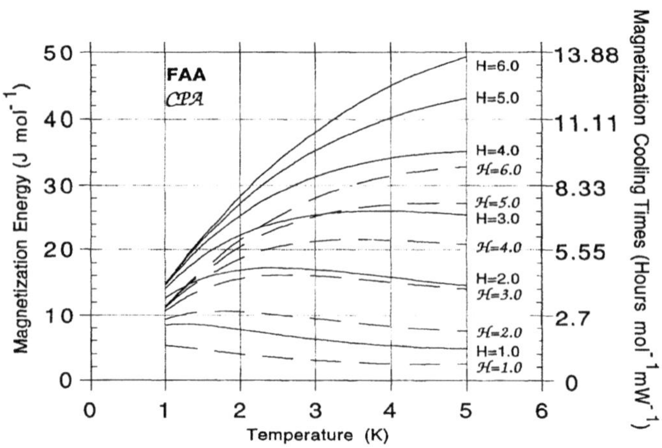

Fig. 1. CPA and FAA magnetisation energy.

atures. The paramagnetic materials for cooling to below 100 mK are usually chromium potassium alum (CPA) or ferric ammonium alum (FAA). The detailed operation of an ADR can be found in most texts on magnetic cooling (*e.g.* White, 1989) and can be summarized in two steps. The first is isothermal magnetisation of the paramagnetic material at the bath temperature, typically pumped ^{4}He, generating heat (Magnetisation Energy) which has to be extracted via a heat switch. The amount of energy generated is shown in Figure 1 as a function of bath temperature and magnetic field for both CPA and FAA. For convenience the right-hand side of Figure 1 has been converted to time for the heat to be extracted at a maximum rate of 1 mW.

The second step can be effected in two ways. One method is adiabatic demagnetisation to zero magnetic field resulting in the paramagnetic material reaching a temperature below that required for the detectors. The detectors are maintained at their operating temperature through the use of a heated detector stage with a weak thermal link to the colder paramagnetic material. This method is referred to as Joule heated detector stage. Once the paramagnetic material has warmed, due to the parasitic heat leak, to a temperature where the detector stage can no longer be maintained then the process starts again with isothermal magnetisation at the bath temperature. The hold time for this method is dependent on the initial paramagnetic material temperature (prior to demagnetisation), magnetic field strength, heated power to the isolated stage and the parasitic thermal load. The second method requires the detector stage to be thermally anchored to the paramagnetic material. The magnetic field is reduced adiabatically to a value corresponding to that required for the detector temperature. The

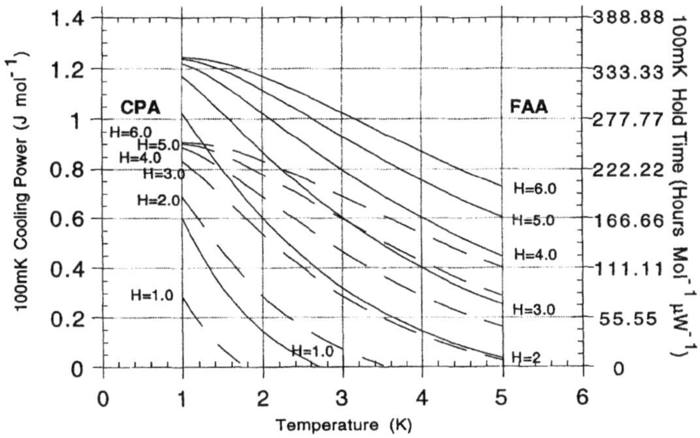

Fig. 2. CPA and FAA 0.1 K hold time.

magnetic field is then reduced isothermally at a rate to counteract the parasitic heat leak, maintaining a constant detector temperature. When no field remains the process starts again with isothermal magnetisation at the bath temperature. This method maximises the low temperature hold time and is dependent on the initial temperature, magnetic field strength and the parasitic thermal load. Figure 2 shows the 0.1 K hold time for a parasitic heat leak of 1 μK for CPA and FAA as a function of initial temperature (bath temperature) and magnetic field for 1 mole of material.

Conventional laboratory ADRs are operated from the lowest possible bath temperature (usually pumped ^{4}He at $\simeq$1.6 K) and the highest possible magnetic field ($\simeq$ 6 T) in order to maximise the low temperature hold time through the reduction of the initial entropy of the paramagnetic material and also the parasitic thermal load. The later may further be reduced by the provision of a secondary cooling stage, of intermediate temperature to the material and bath, attached to the paramagnetic material supports, usually the dominant source of parasitic thermal load, thereby decreasing the thermal load to the paramagnetic material and increasing the low temperature hold time. This secondary stage can either be pumped liquid ^{3}He ($\simeq$ 0.3 K, Ruhl & Dragovan, 1991) or a second paramagnetic material (Hudson, 1972).

2. Adaptation of adiabatic demagnetisation refrigerators for space

If super fluid ^{4}He or solid Hydrogen is to be carried in orbit then very little adaptation of a conventional ADR would be necessary as with the case of the ADRs planned for the AXAF, SIRTF and FIRE spacecraft (Serlemitsos

TABLE I
Current mechanical coolers.

Mechanical Cooler Nominal Temperature (K)	Temperature K	Cooling Power mW
80 K	80 K	800
50-80 K	80 K	1200
	65 K	800
20 K	30 K	300
	20 K	60
4 K	4 K	10
2 K	2.5 K	3

et al., 1992, Timbie et al., 1989, Bock, 1994). However the carrying of cryogen limits the life time and increases the mass of the satellite. Of greater potential would be the coupling of an ADR to a mechanical cooler. However, this puts critical constraints on the ADR forcing a new approach to the magnetic cooling technique. This arises due to the fact that mechanical coolers, unlike liquid or solid cryogen, have limited cooling power which decreases with decreasing temperature. A superconducting magnet no longer immersed in cryogen will have a much greater chance of quenching, causing a rapid temperature rise of the magnet and the mechanical cooler with potentially dangerous consequences for the cooler. The chances of a magnet quench can be minimised by operating with as low as possible magnetic field and the highest possible cooler temperature, for greater cooling power, directly opposite that required for maximum low temperature hold time. Table I details the current space mechanical coolers and their cooling powers at several temperatures (Bradshaw, *private communication*).

A further constraint is imposed on the size of the paramagnetic material due to the physical strength of the material supports. Since the system is intended for launch the material supports must be capable of withstanding the forces involved, thus imposing an upper limit on the mass of paramagnetic material. Any increase in mass, to increase the hold time, has to be offset by an increase in strength of the supports hence increasing the thermal load. This leads to no net increase in low temperature hold time. A mass equivalent to 0.25 Moles of CPA or FAA has been determined to be optimum

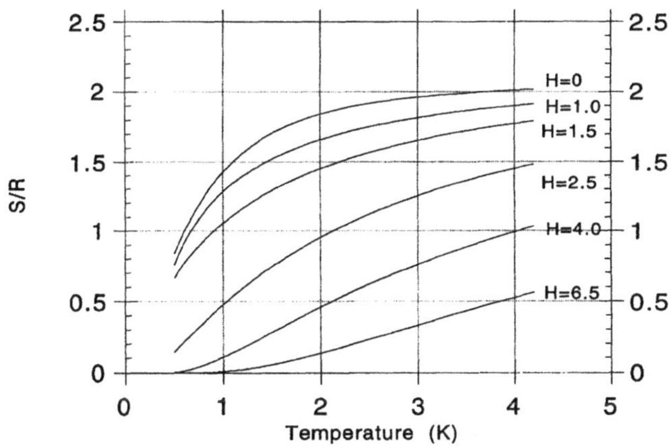

Fig. 3. Entropy diagram for Gadolinium Gallium Garnet.

(Davenport et al., 1992), when the supports are made from Kevlar fibre. An additional contraint arises due to the requirement of the system to undergo bake out since CPA and FAA decompose at 40 and 89 °C respectively.

A mechanically cooled ADR, like its conventional form, requires the lowest possible bath temperature in order to maximise the low temperature hold time. In order to realise a quick recycle time and high efficiency (the ratio of hold time to hold time plus recycle time), a heat sink capable of absorbing many milli-watts of power is required.

Current mechanically cooled superconducting magnets have achieved fields of 1 Tesla (Van der Laan, 1992). Increasing this any further introduces the risk of quenching due to the higher currents and magnetic field. We are therefore constrained at present to consider an ADR cooled by either a 2 or a 4 K cooler with a maximum magnetic field of 1 T. Figure 2 shows that for CPA cooling to 100 mK is impossible for a magnetic field this low however, it would be possible with FAA and a 2 K mechanical cooler operating at 2.5 K.

If the higher temperature tolerant, but lower cooling power, paramagnetic material (CPA) is to be used, a means of cooling the magnetised paramagnetic material to a lower temperature will be required. This can be achieved by using further stages of magnetic cooling provided by the higher temperature paramagnetic materials. One such well known material is Gadolinium Gallium Garnet (GGG). The entropy diagram of GGG is given in Figure 3 and the temperature reached on demagnetisation as a function of bath temperature and magnetic field is shown in Figure 4.

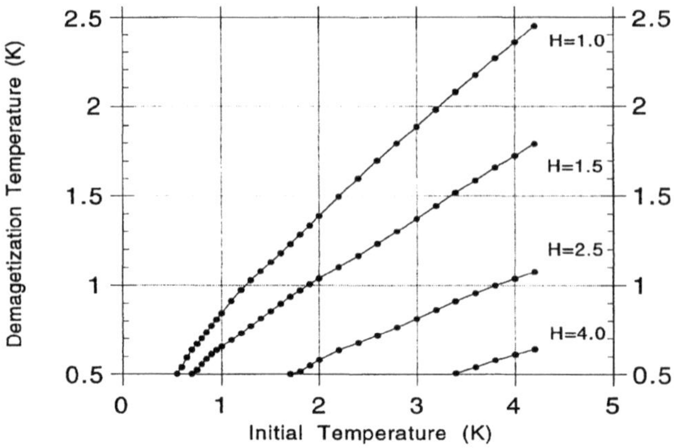

Fig. 4. Temperature obtained on demagnetisation of GGG.

3. A possible ADR for space

In modelling possible ADRs for space we have concentrated on the 4 K cooler since development of this is in a more advanced state than the 2 K version. As can be seen from Figure 4 for 1 T and a 4 K bath the GGG stage will only cool to 2.4 K, not cold enough to enable CPA to be used to obtain 100 mK. We therefore considered an additional GGG stage, making a three stage ADR, providing cooling from 2.4 K to 1.6 K. Figure 5 shows such an ADR composed of 0.25 Moles of CPA and two 0.25 Moles of GGG. The suspension system is made from 20 cords of Kevlar fibre possessing sufficient strength to withstand to a high margin (greater than a factor of 10) the possible forces experienced if such a system were launched. The ADR is operated by sequentially demagnetising the GGG stages to cool the magnetised CPA to 1.6 K. The GGG stages then warm due to the parasitic heat leak. In order to minimize the mechanical complexity, since three heat switches are required, we have considered using gas switches for the 4 K-GGG and GGG-GGG interfaces and a mechanical one for the GGG-CPA interface. Due to the low temperature differences between the gas switch interface, the gas switch parasitic heat leak will be low (6 μW and 1.6 μW respectively) and should not be difficult to construct. Redundant gas switches have been included in the model. We have dynamically modelled this ADR in order to ascertain the 0.1 K hold time, the results of which are detailed in Table II.

Although the hold time for this model is low, it is encouraging since we can see from Figure 1 (CPA cooling power) that a decrease of 0.2 or 0.4 K in the temperature of the second stage, CPA pre-demagnetisation temperature, would increase the hold time by a factor of 2 and 5 respectively, giving hold times of 5.6 and 14 hours. In addition and increase of the magnetic field

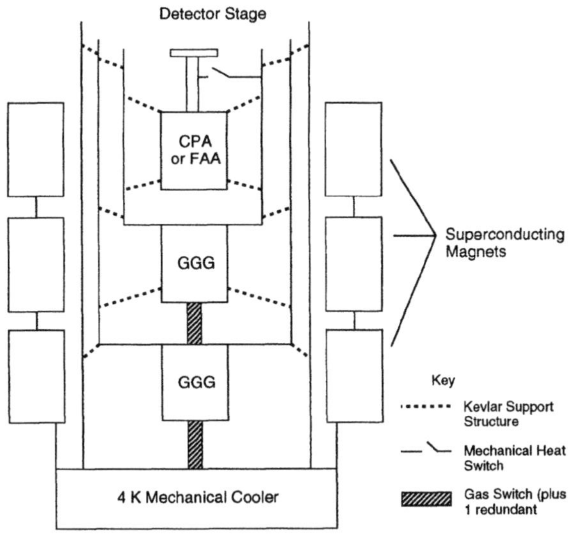

Fig. 5. Three stage ADR.

TABLE II
Dynamic thermal model results.

Magnetic field used	1.0 T
First GGG stage temperature	2.4 K
Second GGG stage temperature (Initial temperature of CPA)	1.6 K
Total initial thermal load on 1st GGG stage	18 μW
Total initial thermal load on 2nd GGG stage	3.6 μW
Total thermal load on CPA	0.7 μW
0.1 K hold time	2.8 hours
Maximum power dissipation	6 mW
Recycle time	0.16 hours
Efficiency	94%
Projected mass	$<$ 10 kg

to 2 T for GGG and CPA would increase the hold time to 19 hours and enable the elimination of one GGG stage, creating a simpler two-stage ADR (GGG+CPA). We further note that GGG at 4 K is not a particularly efficient cooler since for 1 T the reduction in entropy is small. The identification of GGG as a good material was driven by research for obtaining a temperature

of 4 K from a bath temperature of 15 to 20 K using the magnetic cooling technique (Barclay & Steyert, 1982). Research into materials specifically for the 4 to 1 K region has been neglected due to the ability of liquid helium to provide such temperatures more efficiently. We have therefore started a search aimed at specifically finding the best material for the 1 to 4 K range. This is discussed in the next section.

4. Alternative paramagnetic material and future work

Dysprosium Gallium Garnet, (DGG, $Dy_3Ga_5O_{12}$, Tomokiyo et al., 1985) in the 1 to 4 K range is superior to GGG however, it is not as readily available nor as easy to produce. From the entropy diagram given by Tomokiyo et al., a 1 T magnetic field is approximately equivalent to a magnetic field of 3 T for GGG. This material offers the possibility of cooling the CPA stage to 1.0 K and below using just one stage of DGG from a 4 K bath and a magnetic field of 1 T. This would increase the 0.1 K hold time of the CPA to approximately 40 to 80 hours with a recycle time of around 1 to 2 hours. In addition, the use of the 2 K cooler may potentially increase this further. We are at the present time contructing a dynamic thermal model of such a two stage ADR (DGG+CPA) cooled by a 4 and 2 K cooler in order to determine 0.1 K hold time.

5. Conclusion

Through dynamic thermal modelling of an ADR composed of three magnetic cooling stages, made from Gadolinium Gallium Garnet (GGG) and Chromium Potassium alum, a 0.1 K hold time of 3 hours can be obtained with high efficiency from a magnetic field of 1 T and a single 4 K mechanical cooler. We show that if the current magnetic field limit for mechanically cooled magnets is increased to 2 T then the 0.1 K hold time increases to 19 hours and one GGG stage is eliminated. Current work indicates that if the superior Dysprosium Gallium Garnet is used in place of GGG then the 0.1 K hold time may increase to between 40 and 80 hours with only one rather than two stages of cooling.

References

Barclay, J. A., Steyert, W. A.: 1982, *Cryogenics*, **February**, 73.

Bock, J. J.: 1994, *this volume.*

Cunningham C. R., Gear, W. K.: 1990, in *Proc. of the SPIE Symp. on Astronomical Telescopes and Instrumentation for the 21st Century*, Tucson.

Davenport, I., Smith, A., Hepburn, I., Ade. P. A. R., Sumner, T. J.: 1992, in *Proc. of an ESA Symp. on Photon Detectors for Space Instrumentation*, **ESA SP-356**.

Hudson, R. P.: 1972, *Principles and Applications of Magnetic Cooling*, North Holland Pub.

Lesyna L., Roellig, T., Savage, M., Werner, M.: in *Proc. of the 3rd Infrared Detectors Technology Workshop*, NASA Technical Memo 102209.
Rhul, J. Dragovan, M.: 1991, in *Proc. of the 4th International Workshop on low temperature detectors for Neutrinos and Dark Matter*, Oxford.
Serlemitos, A. T., et al.: 1992, *Adv. in Cryogenic Engineering* **37B**, 899
Timbie, P. T., Bernstein, G. M, Richards, P. L.: 1989, *IEEE Trans. on Nuclear Science* **36**, 898
Tomokiyo, A., Yayama, H., Aomine, T., Nishida, M., Sakaguchi, S.: 1985, *Cryogenics* **25**, 271
Van der Laan, M. T. G., Tax, R. B., ten Kate, H. H. J., Van de Klundert, L. J. M.: 1992, *Adv. in Cryogenic Engineering* **37B**, 1517
White G. K.: 'Experimental Techniques in Low-Temperature Physics', 1989, *Monographs on the Physics and Chemistry of Materials*, **43, 3d Ed.**, Oxford Science Pub.

INFRARED DETECTION, REVIEW OF EXISTING AND FUTURE DEVICES

F. SIBILLE
Observatoire de Lyon
F69230 Saint Genis Laval

Abstract. I present the current status of arrays of detectors operating in the 1 - 200 micron range, restricted to mature devices, and two recent concepts of infrared detection which could lead to future developments.

Key words: Infrared – detection

1. Array formats, Pixel size and readout

It is only in the near IR, in the 1 to 5 μm range, that large formats such as 1 K×1 K have been achieved, not far behind the largest silicon CCD's used in the visible. 2 K×2 K formats are even within near reach, either in a single piece, or by grouping of 4 smaller 1 K×1 K items. In the mid or thermal infrared, between 5 and 40 μm, smaller formats like 256 × 256 are readily available, and will soon grow to 512 × 512. For wavelengths longward of 40 μm, only very small arrays have yet been manufactured, relying more on assembly of individual elements or rows than on integrated circuit technology.

With increasing wavelength of operation, the general trend is to decrease the number of pixels, and to increase their physical size. This is due to a fundamental limitation of optics and to a technological problem, which can be illustrated as follows.

On one hand, at least for astronomical applications, a given wavelength implies a minimal pixel size: the beam étendue $E = \frac{\pi}{4}D^2\Omega$, where D is the diameter of the pupil of the optics, and Ω is the solid angle of the pixel field of view. Transfered at the detector level, E becomes $a^2\frac{\pi}{4}(D/F)^2$, where a is the size of the pixel and F/D is the aperture number of the optics. Quality imaging optics require $F/D \geq 1$, and usual applications take Ω of the order of $\frac{\lambda}{D}$, so that finally one must have $a \geq \lambda$.

On the other hand, a large pixel size combined with a large format implies a large integrated circuit, and a small number of pieces manufactured at a time on a single waffer. Manufacturing an array needs a large number of processes, each of them introducing risks of defects, so that the yield of manufacturing arrays of large physical size can be very low.

Except in the very near infrared, these arrays must operate at a rather low temperature, say 30 K and below, where silicon CCD have charge transfert efficiency problems. The most popular readout circuits use an hybrid approach, with the so called DRO scheme (Direct Read Out) which offers the

Space Science Reviews **74**: 225–228, 1995.

advantage of an individual amplifier for each pixel, and a variety of adressing and sampling modes. In particular, for those slow devices operating under low background, the triple correlated sampling mode has proven quite efficient in reducing the readout noise by eliminating both low frequency noises and the kTC noise.

2. Specificity of the different wavelength ranges

2.1. Near IR, 1 to 5 μm range

In the 1 to 5 μm range, mainly intrinsic materials like HgCdTe and InSb, have been favored, because of their high quantum efficiency for a small volume of detector. This, in addition, gives the advantage of a low susceptibility to high energy radiations in space applications.

HgCdTe arrays, with long wavelength cut off near 2.5 μm are produced with high performances under very low background: dark currents as low as 1 electron in 100 s at 65 K, low read noise in the 2-20 electron.rms range (output sampling dependent), and correlatively low integration capacitance of the order of $10^4 - 10^5$ electrons.

InSb has also been widely developed for the 2-5 μm range, although, with proper anti-reflection coating, it can be used in the visible down to 0.4 μm, with still good quantum efficiency. The dark current is higher, typically 100 electron/sec, as well as the read noise, 5-50 electrons.rms, and the integration capacitance, $10^5 - 10^6$ electrons, suitable for the higher background conditions usually met in this wavelength range.

These arrays, operating at typical temperatures of the order of 60 K, are well suited to closed cycle cooling.

2.2. Mid IR, 5 to 40 μm range

Mostly extrinsic doped silicon materials are used in this domain: Si:Ga (bulk, λ_c 17 μm), Si:As (BIB, λ_c 29 μm) and Si:Sb (BIB, λ_c 40 μm). They operate in the 2 to 20 K temperature range, the longer the wavelength cutoff, the lower the required temperature.

BIB detectors (for Blocked Impurity Band, also called IBC, for Impurity Band Conduction) are photoconductors so heavily doped that the impurity energy levels extend in a pseudo conduction band, the undesirable hopping conduction between impurity atoms which would normally result is blocked by a thin transparent layer of pure Si, so that photoconduction can take place only through photoexcitation to the conduction band. The high dopant concentration provides detectors of high quantum efficiency for a much smaller thickness than classical bulk mode detectors, and in parallel, like for intrinsic material, a lower susceptibility to ionizing radiations.

Under low background conditions, as in most space applications, BIB detectors also have the invaluable advantage, over bulk detectors, of presenting much less transient and memory effects with changes of the illumination level.

These arrays have been produced in typical formats of the order of 256×256, with read out circuits optimized for low, medium and high background with respectively integration capacitances of 3×10^7, 10^9 and 3×10^{10} electrons. Typical dark current and read out noise are of the order of 150 electrons/s and 150 electrons.rms, although the high background devices are most often photon noise limited.

Nevertheless, bulk conduction photodetectors present a lower dark current than BIB's, and for this reason could be prefered for specific space applications where the dark current is at stake, and for which there is no large illuminations variations to expect.

2.3. Far IR, 40 to 200 μm range

In the 40 to 200 μm range Ga:Ge is the most developped material, with a long wavelength cutoff at 120 μm for bulk material, 170 μm for BIB and 200 μm for stressed crystal.

Only 1×32 assembly have been made so far, with the aim of building a 32×32 array by putting 32 of them together. Such devices, which require operating temperature about 2 K, presents severe difficulties in manufacturing the readout electronics, usually connected to the detectors by flat cables.

3. Emerging new detection concepts

3.1. Superconducting Tunnel Junctions (STJ)

Although the principle of STJ can be explained in a rather simple way, one should not forget that the detailed physics of the process is quite complicated, and out of the reach of this paper. We shall consider a junction of superconducting Niobium, that is a $Nb/Al_2O_3/Nb$ structure, with an energy gap between the Fermi level and the conduction band of 1 meV. At rest, the electrons are at the Fermi level in Cooper pairs, and do not participate in the electrical conduction. A photon of energy larger than twice the gap, when absorbed in the crystal, can excite a pair in the conduction band, providing two quasiparticles which have enough lifetime to tunnel through the insulating Al_2O_3 barrier, and to be collected in the other part of the junction as a photoelectric current pulse. Actually, if the photon energy is much larger than twice the gap, the quasiparticles will, before tunnelling, cascade down to lower levels in the conduction band, producing phonons, which in turn can have enough energy to excite new Cooper pairs into quasiparticles. The photon detection in such a device has two very interesting features: first, it provides a current amplification, quite useful for photon counting, next, since the charge of the collected current pulse is dependent on the energy of the incident photon, it gives a direct indication on its spectral characteristics, which is a marked advantage over photoconduction. The theoretical upper limit of the spectral resolution is $\lambda/\delta\lambda = (\frac{h\nu}{2\Delta})^{\frac{1}{2}}$ where Δ is the the energy gap width.

The initial developments of this type of detection were applied to the X rays range, but, lately, it has also been tested in the visible, with quite encouraging results, and ongoing works aim at extending its domain into the near infrared.

3.2. Superconducting bolometers

Superconducting materials can be used as bolometer, using the variation of the energy gap width with temperature, which can be sensed with a properly biased diode.

Such detectors present intresting features: they have a wide spectral band pass (150 μm to 3 mm), and they can be manufactured with extremely small thermally active volume. Reported performances are in agreement with the fundamental thermal fluctuation limits.

4. Conclusions

Arrays of large formats and high performances photoconductors, intrinsic and extrinsic, are readily available for applications in the near and mid infrared, while the situation is less advanced, and the technical problems much more difficult, for wavelengths longward of 40 μm.

New concepts of photodetection are emerging, still in an early phase of development, relying on superconducting materials: STJ at shorter wavelengths, with photon counting and spectral resolution capabilities, and bolometers at long wavelengths, with excellent NEP performances. Both present potentialities for manufacturing arrays, but require very low operation temperature, which could be a problem for space missions.

A NOVEL BOLOMETER FOR INFRARED AND MILLIMETER-WAVE ASTROPHYSICS

J. J. BOCK, D. CHEN, P. D. MAUSKOPF and A. E. LANGE
Dept. of Physics, University of California, Berkeley, CA

Abstract.
We are developing a novel bolometer which uses a fine mesh to absorb radiation. The filling factor of the mesh is small, providing a small heat capacity and a low geometric cross-section to cosmic rays. The mesh is patterned from a free-standing silicon nitride membrane and is thermally isolated by long radial legs of silicon nitride. A thin metallic film evaporated on the mesh absorbs radiation by matching the surface impedance to that of free space. A neutron transmutation doped germanium thermistor attached to the center of the mesh detects the temperature increase from absorbed radiation. The low thermal conductivity and heat capacity of silicon nitride provide improved performance in low background applications. We discuss the theoretical limits of the performance of these devices. We have tested a device at 300 mK with an electrical $NEP = 4 \times 10^{-17}$ W Hz$^{-1/2}$ and a time constant $\tau = 40$ ms.

1. Introduction

Bolometers detect infrared and millimeter-wave radiation by converting incident power to thermal energy in an absorber which is weakly coupled to a thermal bath at a temperature T_s. The temperature rise of the absorber is detected by a sensitive thermistor. The thermistor, absorber, electrical leads and supports each contribute to the heat capacity C of the device. The thermal conductance G to the heat sink is determined by the conductance of the electrical leads and the supports. The fundamental limit to the sensitivity of a bolometer is determined by fluctuations in the energy transport between the bolometer and the heat sink, and is given by $NEP = \sqrt{4k_BT^2G}$. Under high backgrounds, the sensitivity of a bolometer is maximized when the temperature of the absorber is raised to a temperature $2T_s$, and then depends only on the temperature of the thermal sink, the optical absorptivity η, and the incident optical power Q,

$$NEP = \gamma\sqrt{4\frac{k_BT_sQ}{\eta}}\ [\mathrm{WHz}^{1/2}],$$

where $\gamma \approx 3$ for our choice of materials and thermistor (Mather 1984). The heat capacity of the device determines the thermal time constant $\tau = C/G$. The responsivity to a modulated signal is reduced by the factor $(1 + \omega^2\tau^2)^{-1/2}$, where $\omega = 2\pi f$, and f is the frequency of the modulation.

At low backgrounds ($Q \ll C\omega T_s$), the sensitivity of a detector is determined by the operating temperature and the thermal conductance. Several practical considerations may set an upper limit on the thermal time constant

Space Science Reviews **74**: 229–235, 1995.

τ, and thus a lower limit on the thermal conductance. Bolometers coupled to conventional readout electronics exhibit $1/f$ noise which requires modulation of the optical signal at several Hz. This in turn requires that the thermal conductivity G be large enough that the reduction in responsivity due to the factor $(1 + \omega^2\tau^2)^{-1/2}$ be not so large as to limit the sensitivity. In addition, sensitivity can be degraded by interactions with cosmic rays which heat the bolometer, requiring several detector time constants for the device to recover.

We are developing novel bolometers in which the cross-section to cosmic rays is greatly reduced relative to the cross-section to millimeter-wave radiation. The bolometers are designed for use in the balloon-borne experiment BOOMERANG (see paper by A. Lange in these proceedings) to reduce the interaction rate with the enhanced cosmic ray flux at Antarctic latitudes. At 300 mK both the heat capacity and thermal conductivity of the bolometers under development are several times smaller than the heat capacity of earlier composite bolometers (Alsop et al., 1992). The architecture of the devices facilitates close matching of the electrical and thermal properties of the detectors, allowing them to be operated in an AC bridge readout with greatly reduced $1/f$ noise (Devlin et al., 1993).

2. "Spider Web" Bolometers

The absorber and thermally isolating supports are patterned from a free-standing membrane of silicon nitride. The geometry of the devices we have analysed and tested was inspired by a spider web, and is shown in Figure 1. The absorber, segmented to form a mesh-like structure with a spacing of $g \approx 160\,\mu$m, is suspended by 16 support legs. A metal film is evaporated on the absorber such that its surface impedance matches that of free space. A 250 μm cube of neutron transmutation doped germanium attached to electrical leads is glued onto the diamond-shaped region at the center. The absorber has a geometric filling factor of 5%, but efficiently absorbs radiation with wavelength $\lambda \gg g$.

The absorptivity of the bolometer can be calculated using a semi-empirical equivalent circuit treatment for lossy inductive grids (Ulrich 1967) by relating the loss in the model to absorption in the metal film. At low frequencies, an electrical impedance of 200 Ω across the device determines the thickness of the metal film for maximum absorption. The calculated absorption for a single pass is shown in Figure 2 as a function of the geometric filling factor of the grid and the ratio g/λ. The bolometer can be placed in an integrating cavity to increase the number of effective passes.

The heat capacity budget of a spider web bolometer is calculated in Table I. At 300 mK the heat capacity of the thermistor and leads dominates over that of the absorber by a factor of $\simeq 30$. Any excess heat capacity asso-

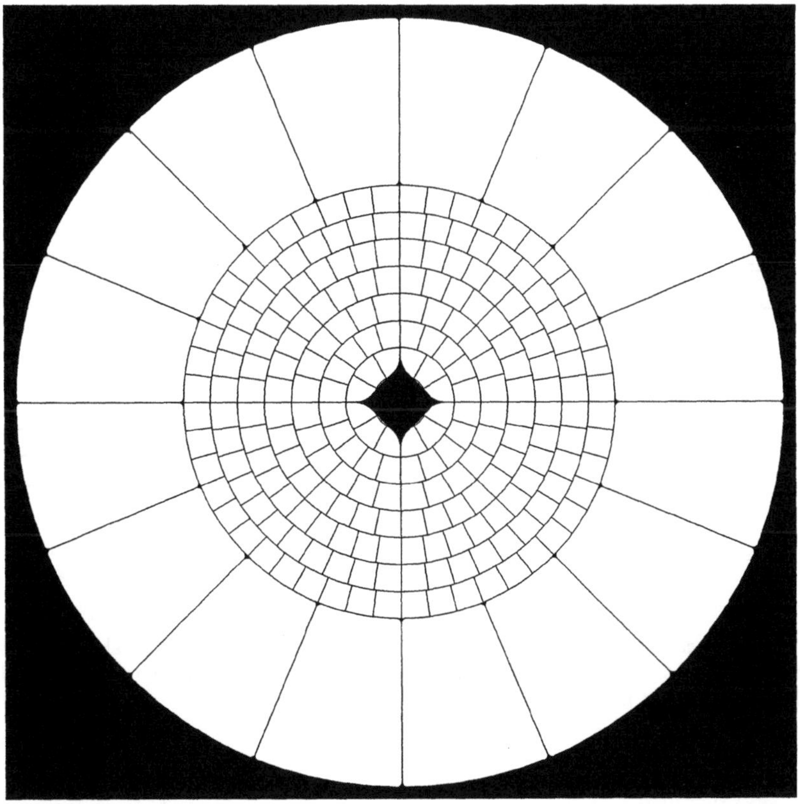

Fig. 1. A schematic of a spider web bolometer which has successfully been manufactured from a $1\,\mu$m thick membrane of silicon nitride. The absorber is 2.5 mm in diameter, the long radial legs are 1 mm long and $5\,\mu$m wide, and the legs in the mesh-like absorber region are $160\,\mu$m long and $4\,\mu$m wide. A metallic film is evaporated in the absorber to match the impedance of free space. A neutron transmutation doped germanium thermistor is attached to the $300\,\mu\text{m} \times 300\,\mu\text{m}$ diamond region at the center.

ciated with surface states or impurities will be greatly reduced in a spider web geometry due to the small volume and surface area of the absorber. The thermal conductivity of the supports given in Table II is taken from a measurement at $T \geq 200\,\text{mK}$ assuming a power law in temperature (Holmes 1994). The thermal conductivity of the supports is small, allowing the thermal conductivity of the device to be dominated by that of the electrical leads for most applications. The thermal conductivity across the absorber G_{abs}, equivalent to the conductivity of a single leg in the central absorbing region, has contributions both from the metal film and the silicon nitride. The thermal conductivity of the metal is related by the Wiedemann-Franz

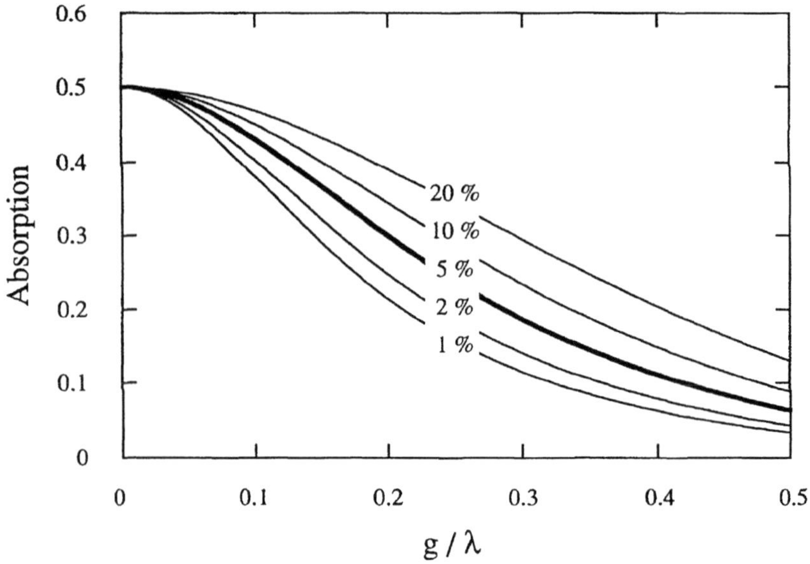

Fig. 2. The calculated absorptivity in a single pass for a lossy mesh. The absorption is plotted against the ratio of the grid constant to the wavelength, g/λ. The relative area filled by the grid is indicated on each curve.

law to the electrical conductivity required for absorption, and dominates over the silicon nitride at low temperatures.

A thermal analysis of the bolometer using a finite element analysis indicates that the thermalization time of the absorber is < 1 ms and that the loss in responsivity due to thermal non-uniformity is $\leq 25\%$. The time required for the absorber to thermalize is of interest because this determines the ultimate speed of the detector. From the thermal model we find that the thermistor equilibrates from a pulse of energy deposited in the absorber over a time $\tau_{\text{therm}} = f\,C_{\text{abs}}/G_{\text{abs}}$, where f is a factor of order unity. This result may be understood by noting that the time for the absorber to equilibrate, neglecting the large heat capacity of the thermistor, is given by $\tau \approx C_{\text{abs}}/G_{\text{abs}}$. As the absorber heats to a temperature $\Delta T = E/C_{\text{abs}}$, where E is the energy of the pulse, it begins to share energy with the thermistor. The time for the thermistor to equilibrate scales as $\tau \approx G_{\text{abs}}^{-1} C_{\text{chip}} C_{\text{abs}} (C_{\text{chip}} + C_{\text{abs}})^{-1} \approx C_{\text{abs}}/G_{\text{abs}}$. The calculated thermalization time is listed in Table II.

Thermal non-uniformity across the absorber will result in a loss in responsivity due to energy transport through the radial supports. We define the effective thermal conductance for power dissipated at the center of the device as G_{ctr}. The effective optical conductance, G_{opt}, is calculated by dissipating power uniformly over the absorber and calculating the thermal rise at the thermistor. For a device dominated by phonon noise, the NEP is degraded by the factor G_{ctr}/G_{opt}. If G_{opt} is optimized for a given background loading, a thermally non-uniform device will be less sensitive compared to a thermally uniform device by the factor $\sqrt{G_{ctr}/G_{opt}}$. The ratio G_{ctr}/G_{opt} is independent of the thermal conductnace of the electrical leads and is determined by the ratio of the conductance of the supports to the conductance of the absorber

$$1 - (G_{ctr}/G_{opt}) = 0.33\,(G_{sup}/G_{abs}); \quad G_{sup}/G_{abs} \ll 1.$$

For the devices listed in the thermal conductance table, $(G_{ctr}/G_{opt}) \geq 0.75$. The thermal uniformity may readily be improved by reducing the A/l aspect ratio of the supports.

We have measured the performance of a spider web bolometer without an evaporated metal film at 300 mK. We obtain a thermal conductance $G = 1\times10^{-10}$ W/K, a heat capacity $C = 6\times10^{-12}$ J/K, an electrical $NEP = 4\times10^{-17}$ W Hz$^{-1/2}$, and a time constant $\tau = 40$ ms. The heat capacity of the silicon nitride membrane alone, determined by first measuring the properties of a thermistor and leads and then measuring the same thermistor and leads attached to a membrane, is $C_{abs} < 1\times10^{-12}$ J/K. The measured heat capacity of the spider web bolometer is 3 times smaller than that of earlier composite bolometers consisting of a diamond substrate coated with bismuth suspended by nylon threads (Alsop et al., 1992). We have measured the effective absorptivity at 90 GHz of a spider web bolometer in an integrating cavity relative to that of a conventional composite bolometer in an integrating cavity. The absorbed power, determined by differencing a 300 K and a 77 K blackbody, is measured for both bolometers, and the effective absorptivity of the spider web bolometer is ≥ 0.7 of the absorptivity of the composite bolometer.

3. Conclusions

We have tested silicon nitride spider web bolometers which have an effective absorptivity similar to that of conventional composite bolometers. The heat capacity of the spider web bolometer is dominated by the thermistor at 300 mK, and the performance may be further improved by reducing the heat capacity of the thermistor. The device shown in Figure 1 approaches the behavior of an isothermal absorber, and may be easily improved by reducing the thermal conductance of the supports.

Spider web bolometers have several advantages over standard composite bolometers. The heat capacity of a spider bolometer is smaller, and unlike earlier composite bolometers, is dominated by the heat capacity of the thermistor and leads instead of the absorber. Larger devices can therefore be manufactured with little cost in heat capacity, allowing coupling to larger

TABLE I
Thermal Capacity Budget of a Spider Web Bolometer

	C electron ($J/cc\,K^2$)	C lattice ($J/cc\,K^4$)	V (cc)	C(300 mK) (J/K)	C(100 mK) (J/K)
Thermistor:					
Au	6.8×10^{-5}	4.5×10^{-5}	5.0×10^{-8}	1.1×10^{-12}	3.4×10^{-13}
Pd	1.1×10^{-3}	1.1×10^{-5}	2.5×10^{-9}	8.3×10^{-13}	2.8×10^{-13}
Ge	2.0×10^{-9}	3.0×10^{-6}	1.6×10^{-5}	1.3×10^{-12}	5.1×10^{-14}
Total				3.2×10^{-12}	6.7×10^{-13}
Electrical Leads:					
Cu	9.8×10^{-5}	6.8×10^{-6}	4.4×10^{-8}	1.3×10^{-12}	4.4×10^{-13}
Nb	-	8.6×10^{-6}	9.5×10^{-7}	2.2×10^{-13}	8.2×10^{-15}
Total				1.5×10^{-12}	4.5×10^{-13}
Absorber:					
Cr	1.9×10^{-4}	1.1×10^{-6}	1.2×10^{-9}	6.9×10^{-14}	2.3×10^{-14}
Au	6.8×10^{-5}	4.5×10^{-5}	2.5×10^{-9}	5.4×10^{-14}	1.7×10^{-14}
Si_3N_4	*	*	2.5×10^{-7}	1.0×10^{-14}	5.0×10^{-16}
Total				1.3×10^{-13}	4.0×10^{-14}

*Extrapolated from Koshchenko et al. (1982).

TABLE II
Thermal Conductance and Properties

	300 mK	100 mK
G_{abs} (W/K)	6.0×10^{-11}	1.4×10^{-11}
$G_{support}$ (W/K)	6.8×10^{-11}	2.5×10^{-12}
τ_{therm} (μs)	250	500
G_{ctr}/G_{opt}	0.75	0.94

throughput with little cost in sensitivity. Further improvements may be realized in reducing the heat capacity of the thermistor and leads. Spider web bolometers have a small cross-section to cosmic rays. Because the silicon nitride supports are mechanically robust and the mass of the absorber is negligible in comparison to the mass of the thermistor, spider web bolometers have greater immunity to microphonics. Lower thermal conductance may be readily achieved by reducing the A/l aspect ratio of the supports. Spider web bolometers are more easily fabricated than standard composite bolometers, and may be suitable for use in low background 2-d arrays.

Acknowledgements

The authors would like to thank P. L. Richards for many useful discussions, W. Holmes for the measurement of the thermal conductivity of silicon nitride, and D. Hebert and the staff at the microlab at U. C. Berkeley for assistance in manufacturing the devices. This research is supported by an NSF/PYI award to Lange.

References

Alsop, D., Inman, C., Lange, A. E., Wilbanks, T.: 1992, 'Design and Construction of High-Sensitivity, Infrared Bolometers for Operation at 300 mK', *Appl. Opt.* **31**, 6610.

Devlin, M., Lange, A. E., Wilbanks, T., Sato, S.: 1993, 'A DC-Coupled, High Sensitivity Bolometric System for the Infrared Telescope in Space', *IEEE Trans. Nuc. Sci.* **40**, 162.

Holmes, W.: 1994, priv. comm.

Koshchenko, V. I., Grinberg, Ya. Kh.: 1982, 'Thermodynamic Properties of Si_3N_4', *Neorganicheskie Materialy* **18**, 1064.

Mather, J.: 1984, 'Bolometers: Ultimate Sensitivity, Optimization, and Amplifier Coupling', *Appl. Opt.* **23**, 584.

Ulrich, R.: 1967, 'Far-Infrared Properties of Metallic Mesh and Its Complementary Structure', *Inf. Physics* **7**, 37.

MIX
Papier aus verantwortungsvollen Quellen
Paper from responsible sources
FSC® C105338

If you have any concerns about our products,
you can contact us on
ProductSafety@springernature.com

In case Publisher is established outside the EU,
the EU authorized representative is:
Springer Nature Customer Service Center GmbH
Europaplatz 3, 69115 Heidelberg, Germany

Printed by Libri Plureos GmbH
in Hamburg, Germany